Sustainable Quantum Bioinformatics

Scrivener Publishing
100 Cummings Center, Suite 541J
Beverly, MA 01915-6106

Publishers at Scrivener
Martin Scrivener (martin@scrivenerpublishing.com)
Phillip Carmical (pcarmical@scrivenerpublishing.com)

Sustainable Quantum Bioinformatics

Edited by

Raghavendra M. Devadas

Preethi

Praveen Gujjar

Sowmya T.

Yashaswini K. A.

and

Bharti Jagwani Motwani

Scrivener Publishing

WILEY

This edition first published 2026 by John Wiley & Sons, Inc., 111 River Street, Hoboken, NJ 07030, USA and Scrivener Publishing LLC, 100 Cummings Center, Suite 541J, Beverly, MA 01915, USA

For more information about Scrivener publications please visit www.scrivenerpublishing.com.

Wiley Global Headquarters
111 River Street, Hoboken, NJ 07030, USA

For details of our global editorial offices, customer services, and more information about Wiley products visit us at www.wiley.com.

The manufacturer's authorized representative according to the EU General Product Safety Regulation is Wiley-VCH GmbH, Boschstr. 12, 69469 Weinheim, Germany, e-mail: Product_Safety@wiley.com.

Library of Congress Cataloging-in-Publication Data

ISBN 9781394444113

Cover image: Generated with AI using Adobe Firefly
Cover design by Russell Richardson

Set in size of 11pt and Minion Pro by Manila Typesetting Company, Makati, Philippines

Printed in the USA

10 9 8 7 6 5 4 3 2 1

Contents

Erasmus of Rotterdam that the term begins to connote a complex of spiritual and cultural values. The retrospective line of descent that the term *Europe* gains with humanism will later be fleshed out fully in Voltaire's formulation of the *république littéraire* (the 'republic of letters'). It is Machiavelli, meanwhile, who first depicts Europe as a land of freedoms in contrast to an Asia in thrall to despots: an image of Europe which will be revived and developed in a very distinctive manner in the eighteenth century in the pronouncements of a learned prince such as Frederick of Prussia and in the perceptive contributions of Montesquieu.

The sorts of complex currents in ideas which wreak qualitative changes upon a geographical area emerge and gain acceptance very slowly. Moreover, the political and the cultural dimensions do not always follow the same path. The all-encompassing insights of scholars can therefore all too easily be dismissed by a political order to which they remain dreams and mirages because of the frontiers, separations and fissures that have persisted in that order until recently and (alas!) continue even today. Bearing these provisos in mind, I shall define my topic and my goals very precisely. I shall recount here the development through the medieval, modern and postmodern periods of an aspect of history which is often ignored: legal history. Legal history is certainly part of more general history, but it has its own autonomous features. The law is sometimes very closely tied to political power, and is often subordinate to it, but, especially in the realities of day-to-day behaviour, and in scholarly reflection, it is often found to have the vigour and the capacity to pursue its own paths.

We will hear the stories of lawmakers, judges, scholars and even businessmen as part of a narrative that is characterized by a continual struggle between the local and individual on the one hand, and the universal on the other. Through this narrative, the law will show itself to be a reality which flourishes at the surface of the everyday, while remaining deeply rooted in the civilization that produces it and therefore capable of fostering genuine insights into that civilization. It is this connection from the everyday to the profound that gives the study of law its independence from the contingent decisions of the political order. And, because the topic of this study is *European law*, it will be guided above all by the dialectical tension between the particular and the universal, between the continent's fragmentation into nation-states and the broad sweep of trans-national legal thinking.

I have used the term *law*, and we must now dwell on this term awhile in order to root out any misunderstandings. In his or her mind, the reader will most probably have identified the idea of *law* with the idea of *legislation*, or written law, and will therefore be unable to comprehend the possibility of the law's independence from the political machinery of government. That being the case, he or she will fail to grasp the full richness of this too often misunderstood concept.

Terminology: Law

It is certainly true that the term *law* harbours serious misunderstandings, especially because, in the European education system and in European culture more generally, the law is given very little attention, meaning that the reader of this book may lack the tools necessary to appreciate it in all its true depth. Particularly to those who live in continental Europe, with the culture of the modern period behind them, the law may appear inextricably linked to power and especially to the supreme power, political power; indeed it may appear little more than an expression of that power. The law is perceived, therefore, as a command from on high, as written law, as an authoritative and authoritarian voice that emanates from the holder of sovereignty.

The law undoubtedly is such a voice. In the context of a modern nation-state, for example, which needs to circumscribe a very complex society, the law can sometimes manifest itself, and perhaps largely does so, as a collection of general legislative acts. However, it would be incorrect and misleading to identify those acts with the phenomenon of law in its entirety. Law consists not only in *power* and *order*, but also in the manner in which society organizes itself in accordance with certain historical values, basing its rules upon these values and observing them in day-to-day life.[1]

The reader should therefore be aware that, even if the law's most visible manifestations are solemn acts of legislation, it is nonetheless part of society and therefore part of life. The law is an expression of society more than of the state. It is the invisible weft that connects the warp of our everyday experiences, permitting the peaceful coexistence of reciprocal freedoms. It is, in effect, society's means of self-preservation.

The social importance of the law is increasingly clear today, as the association between law and nation-state is being brought into question. The conception of the law as identical to the legal systems of nation-states was common amongst previous generations (as we shall see in the second part of this study, Chapter 2), but is now proving inadequate to provide order in a global society such as today's, in which the state and states in general are becoming less and less important as producers of law.

Bearing all this in mind, I shall seek, in the pages that follow, to maintain a broad enough vantage point to command a view of the entire legal landscape. Given that the law is something to be experienced, the legal historian must be the first to refuse to limit himself or herself to mere contemplation of the acts of authority, since these acts do not render a true image of the situation, and a true image is what the historian seeks. I shall not forget that the law belongs not only to the surface of society but, as was mentioned above, is rather *radical* reality – 'radical' etymologically speaking, that is to say it is connected to the roots of society. Nor shall I forget that law is a state of mind more than a set of commandments: the law expresses the values of a society and, by ordering those values, it preserves that society.

Without obscuring the law's connections to political power, I shall therefore dedicate the bulk of my attention to the law that governs the daily life of individuals, what jurists call 'private law'. It is in private law that we can glimpse the vital functions of the law in a living tissue made up of financial transactions, leases, gifts, wills, the acquisition of goods, contracts of employment and enterprises – commercial, agricultural and industrial – in institutions such as these which permit one person to live peacefully alongside another.

The History of the Law as the History of Experiences of the Law

One final clarification is necessary. Our road is a long one – more than fifteen hundred years – and comprises an agglomeration of dates and facts which would risk overwhelming the reader were I not to organize them in a methodologically defensible manner. Do the periods which I examine here – the Middle Ages, modernity and postmodernity – show a uniform understanding of the law, of its origins and its manifestations? That is to say, can we regard these fifteen hundred years as an uninterrupted continuum? Or can we observe differing, even opposed, understandings of the law?

In my view it is clear that we can observe differing understandings of the law in the different periods. Any attempt to identify the diverse and particular shapes which the law takes in the course of the centuries cannot but lead to this conclusion, because it is only by distinguishing between different understandings of the law that we can properly historicize our legal historical material. Any other approach risks burying and smothering the distinctive features of each period under an anti-historical levelling of the landscape.

The most basic methodological precept of this study will be that of distinguishing on the basis of the divergences that I detect in the historical record – divergences that reveal different modes of perceiving, conceiving of and living out the law. It is these differences which reveal to us the different *experiences of the law* in their own manifest and undeniable individuality. My reference to experiences of the law is intended to underline an elementary but often ignored truth: the law is written on the hide of people, it is – as I said a few lines above – a dimension of daily life. The law is inscribed in the concrete facts of life before it is written down in statutes, in international treaties and in works of scholarship. From this broad vantage point we can distinguish the stages of our lengthy route – *Middle Ages*, *modernity* and *postmodernity*. Each stage is circumscribed within very flexible chronological boundaries: from the fourth or fifth century until the fourteenth; from the fourteenth century until the beginning of the twentieth; from the beginning of the twentieth century onwards in a journey that is still under way and whose endpoint we cannot currently see.

Three experiences of the law, three historical civilizations very sharply differentiated juridically which furnish us with three different conceptions of the law and with three very different realizations of it. Ours is not a continuous journey, but rather one made up of three starkly discontinuous episodes – three episodes which present themselves to the historian as three points of maturity to be contemplated and decoded with the greatest respect for their independent foundations. Each of these three points of maturity employs a very particular palette of ideas, of terms and of techniques, and each suffers when these palettes are carelessly confused.

I have set myself here the difficult task of celebrating these differences. I am convinced that this is the only viable way to help the reader along his or her journey of understanding.

1

Medieval Roots

I. A Legal Society Under Construction: The Workshop of Legal Practice

The Political Context: A Society Without a State. The Incompleteness of Medieval Political Power

The first defining feature of the medieval experience of the law, which we will now begin to examine in depth, is its profound discontinuity with the experience that precedes it. Medieval legal thought begins to define itself amongst the strategies and innovations with which the society of the fourth, and especially the fifth, centuries AD sought to reorient itself in the void generated by the collapse of the Roman political structure and of the culture that existed within that structure. Historically, the most salient point is the manner in which the society of the time dealt with that sudden absence of power. For now, we shall deal with the void as it affected the political sphere, which was the most consequential and the most problematic difficulty the new system of law had to face.

A machinery of power as robust, well-constructed and extensive as that of the Roman empire would not, indeed could not, be replaced by one of equal quality and vigour. The novel and defining feature of the era is therefore the *incompleteness* of political power in the medieval period. By incompleteness I mean the lack of any totalizing ambition in the political system of the time: its inability, and its unwillingness, to concern itself with controlling all forms of social behaviour. The political sphere in the Middle Ages governed only certain aspects of interpersonal relationships, leaving others, many others, open to the influence of competing powers.

It is clear that political power – as the supreme power – was exercised in a variety of ways and was often wielded to full effect across certain defined geographical areas. It was also not uncommon to see unlimited power concentrated in the hands of a single prince who used it tyrannically. However, throughout the medieval period, the totalizing and all-encompassing mentality which, as we shall see, will be the distinguishing feature and the ultimate ambition of the princes of modernity is absent. The medieval prince concerns himself only with that which will help him maintain a firm grip on power: the army; public administration; taxes; and repression and coercion of the populace insofar as it helps him maintain order. He is not interested in being a puppeteer who pulls all the strings in the social and economic interactions of his subjects.

We may well ask, and indeed we ought to ask, why this was so: why was political power in the Middle Ages, despite many instances of tyranny, fundamentally weak and above all *incomplete*? The answer is that this situation was brought about by the conjunction of a very particular set of circumstances.

The centuries of transition between late antiquity and the medieval period, that is from around the end of the fourth century until the sixth, bore witness to a great population crisis brought about by war, disease and famine, a crisis which wrought dramatic changes upon the social and agricultural landscape. The population fell significantly and the area of land cultivated fell with it. Subsistence became more and more difficult and the natural world regained its status as an untamed and untameable environment, looming much larger in the collective imagination. The anthropocentric society of Rome, which was founded upon an optimistic faith in man's abilities to subdue nature, was gradually replaced by a more pessimistic attitude with much less belief in man's capacities and far greater emphasis on the primacy of reality. The anthropocentrism of classical civilization was therefore slowly overtaken by a resolute *reicentrism*: a belief in the centrality of the *res* ('thing'), and of the totality of things that make up the cosmos.

This attitude became a collective belief that invested the most insignificant of objects with an aura of power. Power was attributed first and foremost to the natural world, which was seen as a system of primordial rules to be respected. This system of rules conditioned the daily life of medieval communities.

There are also two other, more specific, historical factors which had a great influence on medieval social structures.

One of the defining events of the first centuries of the nascent Middle Ages was the intermingling of the Nordic races with Mediterranean civilization.

Ostrogoths, Visigoths, Vandals, Swabians, Longobards, Burgundians and Franks all established themselves in the Mediterranean region, and built stable socio-political structures there. As one would expect, they brought with them their own political mores, which were distinctive and very different from those they found where they arrived. In the Roman empire an idea of power as sacred, originating in the Orient, had held sway for some time; the holders of power in Rome were therefore seen as earthly manifestations of the divine. The northern races, meanwhile, took a more detached view, seeing power as a practical necessity and casting the wielder of power as his subjects' guide. There therefore arose in the collective imagination a narrative of descent from distant ancestors who were wanderers.

On the other hand, there was the Roman Church, whose influence grew steadily after the fourth century, with an organizational network which spread to the most far-flung territories. Given the absence, or impotence, of imperial power in many of these locations, the Church was by now the *de facto* political power there and could not but frown upon the arrival of a robust rival system, especially one which moved the attitude of the people in an anti-absolutist direction.

The result was, as I have said above, that the political system of the Middle Ages was characterized by a fundamental incompleteness, with important consequences for the rule of law. There certainly was a link from political power to law, that is to say there was law conceived of and promulgated under the influence of politics. This was the sort of law which emanates from on high in the form of commandments; indeed, it was the sort of law to which Europeans were accustomed until recently at the height of modernity. In medieval times, however, such politically generated law was restricted to the areas of legality that were useful to a prince in the exercise of power.

Yet great swathes of the legal relationships which governed the daily lives of the people could not be included amongst these 'political' laws. In these relationships, to which the political system of the times was largely indifferent, the law was able to regain its normal character of reflecting the reciprocal demands of society and the plural currents which circulate through that society. The law, when generated *de bas en haut*, is part of the complex and shifting reality of a society which is in the process of ordering itself and, by so doing, preserving itself. This type of law is not written in the commandments of a prince, in an authoritative text on the paper of the learned; it is an *order* inscribed in things, in physical and social objects, which can be read by the eyes of the humble and translated into rules for living.

An unexpressed but keenly felt suspicion arises that the law, the true law that is, rather than the artifice which helps the powerful maintain their supremacy, is a totality of values underlying social and economic relationships. The law is thus an *order* which functions as a lifejacket for society, whilst the community, aware of this, responds to its values by observing the rules which emanate from them.

Two points must be emphasized.

This type of law is more *organizing* than *empowering* (or *potestative* in technical language). The difference between the two adjectives is not insignificant: the former signifies a bottom-up generation of law that takes objective reality into respectful account; the latter describes the law as the expression of a superior will, which descends top-down and can do violence to objective reality in its arbitrariness and artifice. In a normative vision, law is behaviour itself which, when understood as a value of life in general, is followed and becomes the norm; it is not the voice of power, but rather the expression of the plurality of interests coexisting in any given section of society.

The second fundamental point, and it is one which follows closely from the first, is that, when viewed in this light, the law acquires its own autonomy – despite being submerged in history, and despite being buried under the corporeality of the various interests and fluctuating demands of society. The law emerges as the ordering principle of society, which strives for legal solutions which allow society to continue independently of who wields power. And, contrary to what occurs under the leaden cape of statutory law (in late modernity, for example), where the law becomes the expression of a centralized and centralizing will (*legal monism*), we will observe that the Middle Ages are, throughout, an age of *legal pluralism*. The medieval period demonstrates the possibility of the coexistence of diverse legal orders emanating from diverse social groups, even whilst the sovereignty of one political authority over the territory those groups inhabit remains unquestioned.

It is in this incompleteness of medieval political power, I believe, that the vital key to grasping the 'secret' of the developments in the experience of the law in the early medieval period lies. The distinctive features of medieval law from the beginnings of the era onwards stem directly from this incompleteness.

Given these considerations, the distinctiveness of medieval law imposes upon us certain cultural scruples. We must proceed with extreme caution when deploying vocabulary and concepts closely associated with a modern vision of the law. Indeed, in my opinion we must avoid such terms and ideas for fear of provoking grave misunderstandings. The most problematic of these concepts, although by no means the only one, is the notion of the *state*, which many historians, legal and otherwise, transplant without hesitation to the Middle Ages.

Leaving aside the fact that 'state' could also be used by medieval writers to signify one's rank or social standing, what is most notable for our purposes is that the term *state*, as it is defined and deployed in current usage, has diverged profoundly from the medieval understanding of the term. Indeed, far from signifying a structural continuity, the term has come to denote a concept of extreme historicity: a political entity that is inextricable from the all-encompassing, monopolizing, potestative legal mindset that produced it. In effect, the state is the historical incarnation of political power that has attained perfect *completeness*.

This is not to pose the crude question of whether there was such a thing as the state in medieval Europe, which is the dichotomy to which some have attempted to reduce the methodological problem I am discussing here. Rather, I would argue that, when studying any point in the course of medieval civilization, we should not expect to find the sort of complete political power that we moderns call the state. It is thus an elementary act of intellectual (and terminological) rigour to avoid both the word and the notion *state* when discussing the medieval historical context. That is how I have proceeded in the past and I shall continue in the same vein in this book.[1]

The Triumph of Intermediary Communities: The Completeness of the Community and the Incompleteness of the Individual

This early medieval world – populated by very few inhabitants, scored with perennial political and social disorder, gnawed at by the constant pangs of hunger, lorded over by untamed nature, and afflicted, as we have seen, by a deep-seated lack of faith in the collective – could not help but have a profound effect at an anthropological level, that is to say on the position and role of mankind in the physical and historical world.

One can therefore observe the medieval individual's lack of self-sufficiency and his natural imperfection, his need to bury himself in the bosom of a hospitable and protective community. In a confused and conflict-ridden social reality which lacks the reassurance of a complete political power, the individual has no means of existing peacefully. He will gain it, as we shall see, only with the advent of modernity, when state and individual live in an arrangement of perfect symbiosis and reciprocity.

In the historical context that we are examining here, the incompleteness of power brings with it two consequences that are tightly interlinked. The first is the proliferation of social intermediaries, communitarian groups which take the form of replacements for a supreme power that is absent or deficient. These social intermediaries are necessary organizations in a political reality which lacks solidarity and is therefore incapable of maintaining social equilibrium. The second consequence is that the social intermediaries function as refuges which allow the individual human subject to thrive and to enjoy at least a measure of representation.

Assuming he survives at all, the individual survives as a *socius* ('member' of a society), not as a *singulus* ('individual'); he is a part of a community and not a lone being, defenceless and fragile like an ant outside its anthill or a bee far from its hive. The communities of which the medieval individual was a member vary widely: from nuclei of a few families, to noble houses, as well as guilds, which could be religious, charitable, professional or micro-political. The socio-political reality of the Middle Ages was composed of an extremely fragmented complex of communities, a society made up of societies. This structure would be long-lived and indeed would still be thriving on the eve of the French Revolution.

One further consideration should be added: the powerful influence of that perennial protagonist in medieval culture, the Church of Rome. The Church as a religious denomination was dominated by the idea of a community of the saved, and by a vision of eternal salvation that was problematic for an individual believer to reach in isolation but more permeable to a community possessed of effective sacramental materials, a situation which cannot but have contributed to the tenacious conception of the individual as incomplete and therefore structurally fragile. The sacraments themselves show us the distinction between the Middle Ages and modernity: the medieval communitarian worldview (*extra Ecclesiam nulla salus*, 'no salvation outside the Church') will be replaced by the sacralizing of the direct dialogue between believer and deity. This shift is the hallmark of the quintessentially modern Protestant Reformation.

The Cultural Void and the Factuality of Law. The Primacy of Natural and Economic Facts. The Primordial Facts – Earth, Blood, Time – as Foundational Forces

The void left by the collapse of the public structures of the Roman empire was filled, albeit only partially, by a form of political power which I have described here as 'incomplete'. The fall of Rome also had enormous historical significance for the generation and development of law in the new early medieval society. But another void left by the disappearance of the empire also had a great influence on the development of new experiences of the law, and that was the gap left in the cultural sphere.

The refined Graeco-Roman culture of the previous period left only traces in the closed citadels of early medieval monasteries; it did not circulate at all in wider society. Meanwhile, the West at least appears to have forgotten totally the legal thinking of the ancients. This was a proud cultural edifice of the highest sophistication: built up throughout the republic and the dictatorship to fullness in the age of empire, it forged a perfect symbiosis between the philosophical reflections of the Greeks and the demands of the Roman state. Yet Roman legal thinking was lost by the medievals because it was unusable to them and so remained unused.

To whom might the theoretical niceties contained in the fifty books of Justinian's *Pandecta*, the jewel in the crown of Roman legal scholarship, prove useful? In a socio-economic context like that which took hold from the fourth century on, elegance was not of service; what was needed were tools, albeit rough and uncultured ones, which might help one cope with the gloomy realities of daily life. The greatness of Roman law lay in its academic precision, but the Middle Ages had no space for the deliberations of academics: they sought practical innovations grounded in common sense and pragmatism.

Was this an age of darkness? Do the Middle Ages constitute a time of regression unworthy of historical attention? We should beware of measuring the development of society against a single model. The legal historian, casting an unprejudiced eye over the nascent medieval reality, ought instead to recognize the innovations provoked by the loss of the Roman cultural heritage. Deprived of the inheritance of Rome, and of the undoubted cultural riches that might have been derived from that inheritance, a poorer legal culture had to be built out of procedures which could support and govern that poverty unaided.

It was this absence which led to the construction of an original and novel legal system. If I did not know that total annihilation is foreign to history, I would be tempted to underline the originality of medieval legal choices and solutions to the reader by saying that medieval jurists began again from zero. And where should we look to find the measure of this novelty? There is only one response: in the rediscovery of the *factuality* of law.

Factuality is an unfamiliar and somewhat obscure term; all that is clear about it is its derivation from *fact*. It denotes what happens when the law rediscovers facts in all their force, settles into them and allows itself to be shaped by them, rather than seeking to constrain or alter them. I should make clear that, when I refer to facts, I mean material objects and events, natural features (physical, geological and climatic) and socio-economic phenomena (structures of economic exchange, customs and collective behaviours). When a legal culture is based on scholarship (like Roman law) or on political authority (like modern legal systems), the risk, or the privilege, depending on one's point of view, is that the law is envisaged and devised from on high and projected upon the facts of reality, fitting them, or even forcing them, into its vision. In the medieval context exactly the opposite is true; nature and society are left unmuzzled, whilst the law contents itself with a humble normative role.

It is physical nature especially which masters the law. Nature in the medieval period is a looming primordial force – mysterious, yet alive and fertile, and therefore feared and respected. Medieval man expresses fealty to this force by restricting his behaviour in accordance with the rules he believes he can read in the natural world. The era which we are investigating here seeks its underlying inspiration in a deeply-held *naturalism*: the human dimension is shaped by physical nature, to which it submits docilely. Indeed, so strong is this naturalism as to become a form of *primitivism*. The characteristic feature of the primitivist consciousness consists in an alignment to the natural world so close that person and nature begin to interpenetrate and the boundaries between the two become blurred, until any possibility of contemplating nature critically and objectively becomes lost. Similarly, the nascent medieval socio-legal culture bears witness to a cosmos in which men and things are seen as mere tiles in a mosaic. At the centre of the cosmos are things, not people, especially that great mother-thing, the Earth: an irresistible reality which entrances the human ants whom it nurtures and sustains, but also binds and governs.

The factuality of law in the Middle Ages is important. We shall see that, during the modern period, there is a largely successful attempt to sterilize facts to make them legally irrelevant until an authoritative figure appropriates them and renders them somehow 'legal'. In the Middle Ages, facts are already freighted with potential legal implications that await revelation. Three facts in particular play a determining role in the devising of the new legal order: the Earth, blood and time. The Earth, despite its mysterious vastness, is a maternal figure because it is productive and provides subsistence. Blood links human subjects together indivisibly and spreads amongst them their inheritance of virtue and wealth via means that cannot be communicated outwardly. Time is duration but is also the hammering of months and of years that creates, extinguishes and alters.

These three primordial facts have a single anthropological significance: they reduce the contribution of the individual, elevating nature and the group to protagonist status. The Earth is the resource upon which medieval man may draw to avoid hunger using cultivation and production, yet neither of these processes is carried out by individuals but rather by groups – either families or larger units. These groups reproduce themselves vertically in a chain of successor groups, because only collectively could humanity hope to have any success in the attempt to tame such a mysterious and chaotic reality. Blood is understood as a precious signifier of identity in an ever-broadening circle of allegiances that begins with family and ends with *natio* ('race'), a greater group of individuals who descend from a single stock to make up a single people. Time is understood as a continuing duration and, as such, can only be manifest in the succession of the generations; the individual is therefore effaced as he becomes a mere point in a line. Like memory, time in the Middle Ages is best conceived of by the collective. We can see therefore how Earth, blood and time all emphasize the incompleteness of the individual with respect to the completeness of the community.

To make the historical meaning of this factuality even clearer to the reader, we should emphasize one final consideration: these facts have an immediate impact on the production of laws, because of the legal implications they carry. We shall deal with the Earth in the next section; for now let us say a few words on blood and time.

Blood unites those who belong to the same race but separates these people inexorably from those whom it excludes. Legally speaking, it therefore unites and divides as well. Shared blood means shared rights; different blood means an absolute division under the law. The law is thus reduced to an accessory to one's birth. The principle that is at stake here is what legal historians refer to as *legal personality* – personhood under the law. A venerable legal historiography emphasized the importance of this concept in the medieval period, whilst more recent historiographical studies have tended to reduce its importance in France and in Spain of the period (the Spanish scholars have debated the point particularly ferociously). What is certain is that blood, a primordial fact which distinguishes different races, is of the first importance in early medieval lawmaking. This importance can be seen in the Italian peninsula, a veritable patchwork of legal systems from the fifth century on, in the many so-called *undertakings at law*: the solemn affirmations of the applicability of various legal customs in their specific cases made in front of a judge by a parties or defendants. Undertakings at law were common up to the twelfth century in northern Italy, Lombardy and in the Norman realm (central and southern Italy).

Time is a brute fact in the Middle Ages, a continual accumulation of instants, which can impinge upon the legal sphere simply by the fact of its passing, without any contribution from a human will. This is a very different situation from that which obtained under Roman law, where juridical instruments such as statutes of limitations and positive prescriptions allowed the passage of time to play a role in the loss or acquisition of a legal position, but only when allied with human intervention – an attitude of negligence or diligence on the part of the party concerned.

The Primacy of Custom in Medieval Lawmaking

Let us recapitulate the lines of argument hitherto followed. There have been two guiding principles: *reicentrism* and *communitarianism*. Added to these principles is the widespread medieval tendency to consider the law as a factual entity. This factuality leads to a view of the legal world in the early Middle Ages as one of *custom*, where what is traditional, or customary, begins to generate and solidify new law.

What is custom? A simple but helpful explanatory image is that of a path beaten through a forest. The path does not come into existence until an enterprising subject takes the first steps in a certain direction; he is then followed by a crowd of imitators, all convinced that his is the most rapid way to cross the forest. The path is therefore nothing more than a series of steps, repeated consistently over time.

The same occurs in the formation of custom, which is an action repeated over time in the context of a community, whether small or large. The action is repeated because the members of that community perceive some positive value in it. And the philosophers are therefore correct to define it as a *normative action*: one which, by some peculiar quality, begins to be repeated over a long period of time and becomes the norm.

Since it is an action at root, custom conserves two necessary underlying characteristics. Firstly custom originates from below, from things and from the Earth, from which it cannot be separated; it sticks to the Earth like a serpent and faithfully reproduces the geological, agricultural, economic and ethical structures of the surrounding reality. Secondly, custom originates from the concrete, even if thereafter its significance may be extended by analogy; it therefore carries with it the unavoidable traces of the concrete reality which it seeks to govern with its laws.

But custom is never a solitary behaviour; quite the contrary, it cannot exist without the collective repetition of the action whence it springs. This shows us a further characteristic of custom: it expresses the identity of a group, of a collective – usually a small one at the time the custom is formed, but which can grow far broader with the passage of time. In summary, custom synthesizes the convictions and values that the new legal culture of the Middle Ages placed at its foundations, with the goal of winning its battle with history and guaranteeing its continued survival. It is no surprise therefore that custom shows absolute allegiance to the three fundamental facts of Earth, blood and the passage of time.

To add one further very relevant consideration: because custom originates in a community and expresses its identity, it tends to be projected upon a region until it comes to stand for it and represent the character of its laws. Because of its natural localizing tendencies, custom is one of the factors in the territorialization of the law, albeit a limited form of territorialization restricted to certain areas.

Every region has its own customs; every region forms its own customary laws and moulds them to its purposes. Since custom does not lend its weight to artificial and arbitrary actions but rather to deeper values and convictions, it represents the superficial flourishing of the most profound cultural roots of a given region. Custom is the structure that a place sets up and in it can be seen reflected the deep structure of that place's culture; custom is the structure that allows society to preserve itself when daily socio-political life is often confusing and conflict-ridden. The very rich flowering of customs in early medieval Europe can therefore be seen as a sort of hidden but very solid legal platform. It is in customary law that we may see the *constitution* of the early Middle Ages, deploying the term (which may well sound anachronistic) not in the formal sense that modern jurists use it (a written charter of legal principles, like the Italian Constitution of 1948), but rather as a framework of rules that were not written down but which were nonetheless binding because they draw directly on the values to which medieval society adhered. So the term *constitution* is applicable because custom *constitutes* the various socio-political communities of the Middle Ages, giving each one stability and its own individual shape.

The best place to seek out medieval law is here, in this hidden level of society, rather than in the commandments of the various princes, despite their tyrannical range of powers. Indeed, the princes, as we shall soon see, are required to respect and adhere attentively to custom as much as their subjects are. Princes are not the producers of law: they do not create legal structures, nor does the medieval collective mind identify the dominant trait of their power as being the creation of authoritative norms. The virtue that makes one a prince – that is to say the feature that defines a prince, the ideal to which he has both the power and the duty to adhere – is *aequitas* ('justice'). A prince is a prince because of his ability to dispense justice, a quality which can be derived, in turn, from the lessons written in the tangible world of things and nature. The prince's apparently extensive power thus reveals itself to be an onerous duty.

The Primacy of Legal Practice to a Consideration of Medieval Legal Structures: Under the Banner of Particularism

The prevailing legal landscape of the Middle Ages is made up of a broad framework of the customs discussed above, covering the whole of the European West. This framework has some unifying features, since the aggregate of customs reflects certain structuring principles and underlying choices which are common across the nascent legal context. However, this framework is also extremely fragmented, since each custom is also a reflection of the needs and interests of particular groups or specific local contexts.

The new legal order of the Middle Ages rides under the banner of particularism: that is to say it is an order which cannot and does not wish to smother the demands of the many minorities whom the incompleteness of medieval political power permits to survive and to thrive in all their vitality. Thanks to a fervidly reicentric attitude, the individual realms of the Middle Ages, each with its own legal style and vigour, foster and incorporate thousands of customary laws until these become as defining an attribute of each land as their differing flora.

One might think of a traveller who, when passing from one valley to another, finds that not only the farmland around him has changed but so have the legal customs of his location. Historical sources document this lively diversity, with widespread use of terms such as *consuetudo regionis*, *consuetudo loci*, *consuetudo terrae*, *consuetudo fundi*, *consuetudo casae* (roughly 'local custom/tradition' in each case, with a definition of 'local' ranging from the level of a 'region' to that of a 'household'). These terms appear to show that customs became identified absolutely with their location of origin, to the extent that they begin to stand for and in some way demarcate not only the boundaries between large regions but even those between one homestead and another.

In this context, one can well understand that the vital role of originator of laws is attributed not to a distant and far-off figure such as a prince, but rather to an individual who has the local knowledge necessary to interpret the legal system generated by custom. The law can thus be seen as the means by which medieval man gains his identity and standing within his community.

The protagonist of the medieval experience of the law is therefore not the legislator nor the scholar but the notary: a practical man. The notary is a character who has no truck with legal scholarship; he knows the law only insofar as he learnt it in his professional training and insofar as he needs to know it to satisfy the modest demands of any given case. Drawing heavily on common sense, the notary strives to reconcile the demands of the parties in a matter with the hidden but binding customs of his land.

Silently, unobtrusively, the legal practice of notaries does not create but instead gives concrete form and sufficient technical and juridical heft to procedures which the medieval experience needs in its daily struggle for survival. As we shall see below, it is mainly in the novel field of agricultural law (almost unknown to Roman jurisprudence) that the early medieval notary demonstrates his versatility and makes his positive contribution. Specifically, it is the area of agricultural contracts, which were documents of the highest importance to the medievals, that gives the full measure of the prominence of notaries in medieval lawmaking.

The Marginality of Medieval Legislators

And what of the prince? What of, especially, the many princes who often governed their subjects harshly? Surely from their lofty position these figures ought to be able to promulgate general laws which cut through the heaving mass of customs – all the more so given that they can wield the fearful weapon of coercion? These questions are purely rhetorical, of course: we have already answered them in substance above. But it is worth posing such questions anew since they give us a point of departure to develop further the topic/problem of the incompleteness of power, as I have termed it – a deep-seated cast of mind which shapes legal historical reality.

What seems to us to be the primary and typifying characteristic of the modern potentate – namely the conception of his societal role as that of a legislator first and foremost – is not a perception shared by the early medieval or late medieval collective imaginary. Instead the prince is celebrated by the medieval mindset for his capacities as a judge – as the great bringer of justice to his people. In this he is given great latitude of powers, up to and including the spilling of blood and the say-so over the life and death of his subjects.

Religious, political and philosophical writings of the Middle Ages all emphasize that the greatest virtue required of a prince, and the virtue that most typifies the role, is that of *aequitas* ('justice'). The prince must distribute justice, and specifically he must distribute a form of justice modelled on the world of nature and of things. In his reading and interpretation of the natural world, the prince can be assured of two things: he will find there the instructions for administering truly equitable justice; and he will be able to discover the law, which customs have filtered out of the natural world with the passing of time.

The power of the prince is, and will be for all the duration of medieval jurisprudence, made up of a complex system of powers amongst which judicial authority is central. This system also includes, secondarily, the authority of *ius dicere* ('declaring the law') – the role of making the law manifest to the prince's subjects. Yet, in reality, the prince must come to terms with a constitution fashioned from legal customs which he was not responsible for creating and which, moreover, includes the prince himself under its jurisdiction as much as it does the lowliest of his subjects.

One might object that amongst the primary legal sources of the early Middle Ages, there are a number of legislative texts produced by monarchs or by their chancelleries. This is certainly true: the Visigoth kings in Spain, the Lombards in Italy and the Frankish kings and emperors all produced an appreciable quantity of legislation. But when examined closely, these laws and edicts follow the shape of the wider universe of mores, of customs from time immemorial, which the kings did not dare contravene and to which they submitted.

The medieval monarch shows no creative pride; he limits himself to making manifest in his *lex scripta* ('written law') that which is already contained in the *lex non scripta* ('unwritten law') observed spontaneously by the community. The early medieval attitude towards the term *lex* ('law') is very particular: the conceptual gap that separates *lex* and *consuetudo* ('custom') in modern formalist legal thinking is entirely absent. A *consuetudo* is merely a law that has yet to be made, and a law is merely a custom that has been properly written down, certified and codified.

One final consideration should be noted with regard to the content of legal sources. Often these sources are compendia: haphazard selections of specific or occasional legislation; collections of laws enacted by previous monarchs; edited versions of legal texts from the late Roman empire. They often take as their subject topics of general public importance connected with the exercise of supreme political power over the territory in question. One Italian example would be the penal code and family law contained in the extensive *Edictum Langobardorum* promulgated by the Rotharian kings in 643. The Frankish kings and emperors, meanwhile, produced the *Capitularia*: a series of legislative acts put forward during the long reign of Charlemagne in the late eighth and early ninth centuries. The *Capitularia* are mainly concerned with the rules governing public administration and the relationship between temporal and ecclesiastical power.

Legal Solutions for Daily Life in an Agrarian Society

The very general discussions entered into thus far might run the risk of seeming generic to the uninitiated reader. I am also conscious of the need to keep the promise I made at the beginning of this study to write about the law as a mentality, which would seem to demand a focus on the dealings of individuals. I shall now turn, therefore, to everyday life in the Middle Ages, to the solutions and choices which govern it, so as to fill out our picture of medieval law and give it concreteness. I shall examine here the contractual relationships between individual subjects in the languid dynamic of the relatively stagnant medieval economy. Furthermore, I shall look at the legal procedures used to regulate the relationship between people and goods, especially between people and the land – an essential aspect of an agrarian civilization. To lend weight to my conclusions, I shall make use of the comparative method, with its superlative ability to highlight discontinuities. In this case the discontinuities we will find exist between the medieval context and the choices made by Roman lawyers, whose first aim is always to maintain control of the legal sphere, which Roman jurists saw as vital to the stability of the political order. The ancient Roman approach is, of course, antithetical to the one we found in our examination of medieval legal thinking. I shall therefore give what I hope will be a helpful comparative evaluation of the two systems.

The Romans ensured their contractual agreements were governed by a rigorous principle of standardization: private citizens were allowed standard types of contracts, which followed pre-made models. These offered legal protection only to certain established types of transaction: sales, rentals, loans, etc.

The early medieval practice of contractual negotiation, on the other hand, is characterized by a total lack of standardization. Indeed things could hardly be otherwise in such a profoundly custom-governed legal environment. Custom abhors a rigid template – its moulds are malleable and mutable. Instead it places total faith in the instincts of the notary and in the good faith of the parties. Often the will of parties to a medieval agreement was only free in theory, since they were bound by the demands of the pervasive network of customs. They therefore submitted to the types of undertaking which custom defined for them and towards which the notary would not hesitate to point them. The contractual models shaped by notarial practice functioned as very flexible formal receptacles; they differed dramatically from place to place, but all were able to accommodate a wide variety of customary content.

Roman law is similarly rigorous when it comes to the jealous territory of the relationships between people and property, especially those between people and the land. Roman civilization across all the centuries of its development was always a fiercely proprietary society: that is to say it was founded on the ideal of individual private property. The importance of private property exceeded the purely economic realm and became of political importance as well. The Roman legal term for so-called *real property* rights, that is the rights of people to own *res* ('things'), is *dominium. Dominium* describes a right of property which binds a thing tightly to its owner; the owner's independence and freedom to enjoy his property is respected and safeguarded by the law, which guarantees his *potestative* ('discretionary') rights over the object in question.

The Roman state was careful to institute a system known as *fundamentum rei publicae* ('foundations of the commonwealth'), which carefully circumscribes the conditions on the ownership of property enforced by coexistence with other citizens. This system renders any impingements in fact by other parties on a citizen's property (such as physical contact, use, enjoyment) of no legal standing. There may be physical contact, use or enjoyment of the object in question by parties who do not own it, but these have no legal relevance and do not compromise the integrity of the proprietor's ownership. Hence the distinction between *dominium* and *detentatio*, that is to say between full ownership and discretion over the object as guaranteed and safeguarded by the law, and the simple having of a thing in one's possession. With a piece of land, for example, the anthropocentric culture of Rome would emphasize rights of the holder of the proprietary title, even if he had never set foot upon his land. The owner's rights would be upheld over the rights of those without titular claims to the land but who were nonetheless intimately acquainted with it. A tenant farmer who rented his land from the owner is thus considered a mere *detentator* ('occupier').

Obviously, this entire legal framework was abandoned by the new reicentric culture of the Middle Ages, which did not believe in a puppeteer pulling the strings of the legal order, and depended instead on custom and factuality. Economic facts such as use, enjoyment, trade, or even the simple material fact of physical familiarity with an object, leave the hinterland of legal irrelevance and take up their own place and significance in the eyes of the law. This occurs especially once the passage of time has rendered these situations of fact *effective*. As we have seen, the medieval constitution is not concerned with *validity* – that is the compliance with an authoritative general principle – so much as with *effectiveness*.

To this should be added the change in outlook towards one's physical and social context characteristic of the Middle Ages. The medieval world is no longer seen from the point of view of the subject but rather from that of the object, with the result that the world is understood from the ground up. Things are no longer constituent parts of a landscape bestridden by the autonomous subject, but a living reality whose objective demands the subject must interpret and respect. This impression is entrenched all the more by the necessary facts of cultivation and production required for the subsistence of the community.

Roman law is primarily a civil law, and is therefore predicated upon a legal party who is abstract and economically undefined: the *civis* ('citizen'). Early medieval law, meanwhile, is predominantly an *agrarian law*: it is predicated on the fundamental economic facts of cultivation and production and on legal parties who are assumed to be growers, breeders, woodsmen or suchlike. Medieval law is not governed by the cult of proprietary titles (although this certainly does survive), but seeks rather a more abundant and higher-quality agricultural harvest, in whose name all sorts of sacrifices may be demanded, even from registered titleholders of property.

From a technical legal point of view, there are several conclusions that readily suggest themselves. The medieval legal system favours procedures that provide effective resolutions with regard to land, particularly where agricultural activity is involved. The Roman opposition between owner and occupier appears not to obtain in the medieval period. Many occupiers of land under licence – particularly those who seek to improve the land's productivity in the long term – gain a status of para-ownership thanks to an unobtrusive but continuous erosion of formal property rights. The practice of lawyers in the early Middle Ages, although rough at the edges and lacking in technical sophistication, is already making advances which, in the late Middle Ages, will be formalized into a fully rounded body of legal thought.

The Church of Rome during the First Millennium: The Making and Formalizing of Canon Law

The Church of Rome is the pre-eminent figure at every level of medieval culture: religious, cultural, socio-economic, political and legal. Indeed, one could say that medieval culture is, for the most part, a creation of the Church.

The history of the Roman Catholic Church is of particular interest to the legal historian because it is the only religious denomination which takes it upon itself to create its own original body of law, drawing its authority directly from that of Christ as divine legislator, rather than from any temporal political system. This body of law develops into a unique legal system: canon law. Canon law is by no means the discipline of an isolated priestly caste: in a historical context such as that of the Middle Ages, where Heaven and Earth meet, sacred and secular intermingle, and the citizen and the believer join in one complete unity, canon law cannot but be integrated into the medieval legal order and, indeed, it makes a significant contribution to the shape of that order as we find it.

The reasons behind the Church's decision to enter the field of law are complicated, but I should attempt to give a conclusive answer. It is clear on the one hand that canon law represents a reaction to the Church's need to forge for itself effective instruments of power and influence. Nonetheless, the main reason for the existence of canon law, in my view, is anthropological: in order to obtain salvation, there was a need for a society of the faithful – i.e. a structured hierarchy comprising the Church and its community of believers. Because the individual believer needed a social structure in which to find his place, there was therefore a requirement for a system of laws to govern the Christian community.

And so the Church of Rome, as a legal entity, is concerned from its beginnings to formulate a system of laws suitable for its governance. The first millennium of the Church's existence is scarred by many heretical movements, and its efforts are above all directed towards solidifying religious orthodoxy into stable theological truths. Nevertheless, over the course of the first thousand years of Catholicism, canon law makes slow but steady progress until it assumes a definable shape. Because canon law develops over so many centuries, and is produced in the most distant reaches of contemporary Christendom by a very diverse series of authors (popes, councils, bishops, religious orders, customs, theologians, jurists, etc.), it is unsurprising that the laws of the Church at first grew into a confused morass of rules, many of them inter-contradictory. The situation became an embarrassment for an organization dedicated to a mission of general salvation.

At the end of the first millennium the negative aspects of the canon law of that time had become glaringly apparent. Fortunately, there emerged some far-sighted jurists who began a robust campaign of putting the enormous quantity of material in order: consolidating some parts and harmonizing these with others. We should remember in particular the work of one French prelate: Ivo, Bishop of Chartres. At the end of the eleventh century – during the period known as the Gregorian era after the dominant personality of the time, the centralizing pope Gregory VII – Ivo succeeded in systematizing completely the canon law, producing a careful, unstrained interpretation of all its idiosyncrasies.

Ivo catalogued the many discrepancies and contradictions (*discordantiae*) that had accumulated over the centuries. In an important move for canon law's pastoral ambitions, Ivo resolved the problem by identifying two separate and dichotomous levels of meaning in Christian legal texts. First is that of divine law (*ius divinum*): perpetual and universal law which stems directly from God and is composed of a few essential rules (do not kill, for example). Divine law is immutable because it is vital to every human soul on the path towards salvation. Below divine law comes human law (*ius humanum*), which originates from the Church, from jurists and from custom. This level of law makes up the great mass of canon law and is merely useful for salvation, rather than essential. Since it is only useful, human law must accommodate itself to human frailties, taking into account such variables as differences of place and time, and the circumstances and motivations of actions.

Ivo's reordering of the law gave birth to a legal system which, while it certainly was not compact, now possessed a restricted core of extremely solid primary rules, surrounded by a much broader, more fluid periphery of secondary rules. The secondary rules, of course, needed fluidity in order to be able to find equitable ways to accommodate all of the differing circumstances into which the earthly pilgrim might wander. Moreover, canon law was naturally available to any medieval executor of rules, particularly judges, who, in their concrete evaluations of given situations, could apply either stringency or leniency in order to arrive at the outcome that best furthered individuals' salvation. Such leniency could even extend to a total non-application of the law if required – something that the canon lawyers called *relaxatio legis* ('relaxing the law').

Ivo did not invent any laws; he merely applied a general and longstanding principle of the Church's legal tradition, that of *aequitas canonica* ('canonical justice'), which called for the adjudicator to consider the specific actions of the individual believer and the circumstances in which these had occurred. In so doing, Ivo succeeded in putting forward an accurate interpretation of the canon law which took account of its ultimately pastoral nature. For this reason, the division made by this eleventh-century bishop from Chartres between *ius divinum* and *ius humanum* has stood the test of time and is still considered valid to this day by the Roman Catholic Church when interpreting its laws. From a legal historical point of view, we should highlight one further conclusion: the dominant influence of the Church of Rome and of its legal system in the Middle Ages means that the flexibility of human canon law becomes representative of the entire medieval legal process.

II. Medieval Maturity: The Laboratory of Learning

The Turn of the Eleventh and Twelfth Centuries: A Watershed in History. Socio-economic and Cultural Contexts

At the end of the eleventh century the substantial changes which time had unobtrusively but continuously wrought became more obvious. It is therefore justifiable to see the decades which straddle the division between the eleventh and twelfth centuries as a boundary between one historical moment and another, very different, one.

The agricultural landscape has now changed: where before it was a mixture of woodland and pastureland, now the countryside of Europe has been deforested, its clods broken up and reclaimed for agriculture. The number of inhabitants living on that land has also recovered.

The collective consciousness also appears transformed: the former wariness which forced people to seek the security of a castle or a walled town is being gradually but definitively replaced by a more widespread attitude of trust and confidence. The signs of this change can be seen in the greater circulation of individuals around the continent and the progressive repopulation of the cities. The landscape of Europe is also growing more complex: although the rural sector remains dominant, the cities are growing in importance. And the city itself is far more than a collection of stones: it is above all a spiritual achievement. Whilst a castle may be set apart at the top of a hill, a city will be located at the intersection of the great trading routes which have now been established. The late medieval city does not exist in a state of autarchy, but is rather founded on a desire to be open to outsiders: it thrives on the contributions of people and goods from beyond its walls. This gradual but growing importance of cities demonstrates the renewed confidence late medieval man had in broader social relationships with those from further afield.

The range of socially significant roles also broadens with the rise of the professional merchant. The old markets of early medieval Europe, where local producers traded local goods, will no longer suffice. Given the greater abundance of goods for supply in the late Middle Ages, there is greater demand for long-distance trade. The importance of currency as an intermediary also grows, therefore: a further testament to the greater economic vitality of the period and the stronger bonds of confidence between individuals. A new historical personage arises: the professional merchant who resides in a city and relies on the whole of Europe as his trading space. The merchant himself is, of course, an indication of increased trust between people and peoples; he signals an openness to ever wider socio-economic horizons.

So much for the socio-economic aspects of the new historical context. What of culture? The early Middle Ages possessed plenty of schools and centres of great learning which carried out profound investigations of a theological or philosophical nature. But this knowledge tended to be confined to the monastery; it did not permeate the institutional walls to enrich early medieval civil society. In the late Middle Ages, however, schools began to appear more often in the centre of cities, attached to the cathedral. Cultural learning could now start to circulate more widely.

The cultural void I talked about earlier, which led me to speak of naturalism, and even of primitivism, began to be filled. This is demonstrated by the twelfth-century renaissance: a renaissance created not by the musings of isolated figures but by large personalities, who existed within a cultural matrix that covered all of Europe, and who engaged in lively debate with their peers. The schools which took it upon themselves to foster these debates were by their nature opposed to the stifling influence of particularism, and strove instead for universal ideas. Thus began the great and, before long, widespread trend of founding universities.

We shall deal here only with the very profound realignments which occurred in the fields of theology and of philosophy, since it is from these disciplines that the study of law derived the intellectual nourishment which allowed it to start flourishing in such fertile soil.

Political Power and the Law: The Marginality of Late Medieval Legislators

One can identify a number of innovations, occurring around the end of the eleventh century and the beginning of the twelfth, that affected the underlying structures of Western culture and the collective consciousness. Men of culture began to circulate with the merchants along the reopened arteries of communication. They brought with them the basics of scholarly reflection, and so filled at least one of the two voids which we found to be determining for the shape of early medieval civilization: that of the sudden disappearance of ancient Roman culture.

At this point the reader might legitimately ask what effect these apparently significant changes had upon experiences of the law in the medieval period. Is it correct to categorize together experiences of the law across all of the period between the fall of Rome and the fifteenth century, or does the watershed I have just described suggest a more deep-rooted separation between the early and late medieval periods?

This is not only a legitimate question but a very apposite one, since the answer is, as my introductory comments suggested, that experiences of the law took on a unified shape throughout the Middle Ages: they appeared similar throughout; their foundations remained the same; and the approaches to living out and understanding the law remained coherent. One proviso must be attached to this conclusion, however: the historical period which the modern era so disparages with its epithet *medieval* (literally 'middle period', i.e. 'time of transition'), lasts for the best part of a millennium. The large amount of time to which the term *Middle Ages* therefore refers cannot but imply some variations in the historical experiences of the law during that period, since time never passes without leaving its mark. Nonetheless, we should not be deceived by this: the fundamental choices of medieval legal thought remain substantially constant.

Our task, and it is no easy one, is to justify these apparently generic and unsatisfactory assertions. We shall start right away by pointing out that despite the fact that the cultural innovations of the late Middle Ages were far from insignificant, they did not fracture the pre-existing medieval socio-cultural identity. Instead they were integrated harmoniously into the law's fact-based approach, which had been established through centuries of practice. In so doing, the cultural changes of the late Middle Ages lent medieval legal practice the further vigour necessary to satisfy the demands of a more dynamic, complex society, which was simultaneously rural and urban, agrarian and mercantile.

There is one point that must be emphasized: although the cultural void has been filled, the political void remains just as gaping. The kind of intrusive government which believes itself able and entitled to intervene at a social level and to control the legal dimension of its subjects' lives by producing all the laws which govern them finds no place in the Middle Ages and will not come about until a later period.

The prince continues to be thought of in the collective consciousness as the supreme judge of the community, with one fundamental, non-negotiable quality and virtue, that of justice: the ability to make equitable decisions based on the true nature of things. John of Salisbury, an English prelate who, in the mid twelfth century, wrote the first great tractate on political thought of the Middle Ages' era of learning, depicts the prince as an *imago aequitatis* ('image of justice').[2] The law, meanwhile, John calls the *aequitatis interpres* (the 'interpreter of justice'), foreshadowing the description of it a century or so later by Thomas Aquinas, the consolidator of most medieval theological and philosophical certainties, as *custos iusti* (the 'guardian of that which is just').[3] In other words, the prince is not seen as a supreme will, with arbitrary power over his subjects, but rather as playing a role of attentiveness to nature, the great text in which the lessons of justice are written. This is why St Thomas himself, in his definition of *lex* ('law'), identifies it as a product of reason and thought: the law is not used to project a despotic will upon a community of subjects, but rather to keep that community in order (it is a 'reasoned structure directed towards the common good').

The collective consciousness still does not think of the prince as a *legislator* – that is as a maker of laws. His duty of reading the text of nature will not produce universal and authoritative principles but will rather set the specific parameters of true justice. Indeed the prince himself does not see the *legislative* function as the defining characteristic of his power. The relative indifference to the law which we observed in the early medieval period continues in the later Middle Ages, as does the relative lack of a coherent programme of legislation. The prince is limited to producing such rules as govern the limited field of the exercise of public power. And it is obvious that this should be so, since the dominance of custom continues, meaning that the law retains the imprint of custom, and political power is therefore marginalized and rendered subaltern as a source of lawmaking. This picture certainly holds true in central and northern Italy, where there is an upsurge in civic sentiment that is made concrete in the rise of city-states. The statutes of these city-states function as an expression of their autonomy and are usually fairly loosely drafted with no pretence to comprehensiveness: much more attention is paid to small-scale problems of town planning than to any great institutions intended to regulate citizens' lives. The situation is no different in southern Italy, where one of the great medieval monarchs, Fredrick II, had drawn up for the Kingdom of Sicily a magnificent, sprawling legal monument: the *Liber constitutionum regni Siciliae*, known as the *Liber Augustalis* – an ambivalent text which mixes numerous dated political ideas with a few revolutionary insights.

North of the Alps, in the final full century of the medieval period – the thirteenth – we find some eloquent confirmations of trends discussed here. The German-speaking lands continue with government by customary law for the whole of the century in question, but the monarchies of France, Spain and Portugal are beginning to move down a path that will lead in the end to their development into recognizable nation-states.

Not long after the thirteenth century, France will become the true testing ground for the politico-legal framework of modernity. The French monarchy, still heavily conditioned in its actions by the legacy of feudalism, began to create for itself in the 1200s a more defined and broader political space in which to operate. The greatest innovator in this effort was a king of undoubted managerial abilities: Philip II Augustus (1180–1223). During the course of the century the legislative activity is, however, sporadic and limited to topical interventions. The daily life of citizens in peacetime continues to be regulated by the age-old framework of custom, whilst the king is above all the 'guardian of custom', as Beaumanoir, one of the greatest French jurists of that century, puts it.[4] The first great reformist *ordonnance* (*ordonnance*, or 'ordinance', was becoming the normal term for general laws made by the king), was issued by St Louis (King Louis IX) in 1254 on his return from the Seventh Crusade. However, this edict is directed at royal administrators, and the king does nothing more than reiterate the validity of local customs.

In Portugal royal legislation does not become significant until the reign of Alfonso III (1248–79).

In Spain, Catalonia, Aragon, Valencia and Navarre are all dominated by local customary legal systems until the middle of the thirteenth century.

In the second half of that century, in Castile, Alfonso the Wise (1265–84) introduces an important piece of legislation called *Las siete partidas* ('The Seven Headings') – a very distinctive work which sits somewhat unhappily with Spanish law's localism and dependence on custom but which speaks volumes for the king's abilities as a legislator. *Las siete partidas*, named for the number of its internal divisions, contains mainly universal laws rather than Castilian ones. It therefore draws heavily on Roman and canon law, whose importance in the late Middle Ages we have yet to discuss but which were, it will suffice to say, still principally academic disciplines. And, indeed, the practising lawyers of Castile decisively rejected Alfonso's legal masterwork as alien, despite the fact that it was written in their own familiar tongue.

Western Society's Rediscovery of Complexity Requires New Legal Methods

The continuing dominance of customary lawmaking into the mid thirteenth century, and with it the persistence of an emphasis on factuality in the law, serve as ample evidence of continuity between the legal systems of early and late Middle Ages. This continuity is not only substantial but, I would argue, defining.

Custom is a friendly, nurturing source from which to generate law: it respects local differences and local needs. Nonetheless, custom has one intrinsic defect which I have been at pains to emphasize: fragmentation – it cannot but express a particular set of circumstances. In a less complex social order like that of the early Middle Ages, when society was relatively static and social change occurred at a leisurely pace, custom was perfectly capable of fulfilling the role of the sole legal framework which governed that society. However, custom's innate tendency towards fragmentation meant that it became unsuitable as the sole generator of law when the social, economic and legal landscape became more developed – especially when economic relationships begin to carry a similar weight to legal ones. The Crusades ensured that these relationships were knit together into a social fabric that extended from the Hanseatic ports of the Baltic to the Mediterranean Sea.

In such a complex and diffuse political environment, custom reveals itself to be an unsatisfactory ordering principle. It was clear that facts and customs must remain the primary determinants of the law, but when those facts and customs were spread out across a very large geographical area, and when the needs of agriculture had to be balanced against those of a vibrant mercantile economy, a need arose for broader schemata, more general organizing categories and more rigorous and refined legal approaches than custom's universe of facts was able to provide. What was needed was an ordering framework into which to fit the facts of custom: one which would not stifle them but which would rather organize and systematize them. There was a need to bring some unity to the diversities of custom, since otherwise unmitigated chaos would reign.

There were two sources of law suitable to achieve this aim: legislation and scholarship. These were two sources of law that might lay themselves over the mass of facts and particulars and organize them according to principles, ideas and general patterns. A prince, whether a monarch or the head of a city-state, might very well perform such an operation, but this would involve renouncing his duty to adhere to nature and facts and turning instead to the setting of rules. Princes are still not allowed the role of legislator in the late Middle Ages. Instead only one option remains to a medieval culture that has by now rediscovered the importance of learning: that of scholarship, legal scholarship, to be precise.

The Role of Legal Scholarship. Particularism and Universalism. Customs and Scholarship in the Late Medieval Legal System

The late Middle Ages could well be called an age of learning since scholarship, legal scholarship in our case, takes up a primary cultural role. In a very significant development, medieval legal systems begin to allow scholarship to design the laws which their historical moment so greatly needs.

There are many reasons behind this rise of scholarship, the first of which was highlighted in the previous section: scholarship was the only source which, in the absence of a comprehensive political system, could gather together and organize a huge and disparate body of factual material. Only scholarship could make facts into the sort of ordering principle which any system of law requires by definition. This was a sizeable advantage, since the theoretical categories and principles to which scholarly reflection gives rise are by their nature elastic and therefore well suited to a legal system in a continuous state of development, whilst the authoritative pronouncements of a prince are necessarily more rigid when translated into general commands. As I argued above, scholarship organizes ideas not by suppressing points of difference, but by incorporating those differences as points of nuance to the broader sweep of lines of argument.

The profound importance of legal scholarship in the late Middle Ages is also predicated on other solid reasons. The general editor of this Making of Europe series, Jacques Le Goff, wrote a seminal volume entitled *The Birth of Europe*.[5] The title of the Italian edition this book was *Il cielo sceso in terra* ('Heaven Descends to Earth'), a choice which reflects one of the great medievalist's most valuable insights: the extent to which medieval civilization was a historical moment totally focused on a reality beyond nature and history; the people of the Middle Ages certainly lived in space and time, but they saw their ultimate resting place as existing beyond those spatial and temporal boundaries. In such a context the jurist or master of laws, in his role as a learned man, is more intimately connected to God as the ultimate wisdom and ultimate truth than any other earthly worker. He is seen as an enlightened and enlightening being, a sort of mediator between heaven and earth, placed on a higher plane than any other searcher after truth. This elevated position explains the medievals' great faith in their jurists.

However, so as not to jump to misleading conclusions, we must add that this scholarship is of a concrete, pragmatic nature. As we shall see, medieval jurists are no cloistered academics, foolishly absorbed in theoretical projects entirely abstract from their context. Instead, this is an age of great thinkers – mostly teachers at the many universities now dotted across Europe[6] – real flesh-and-blood characters, well integrated into civil society and often occupying positions of power and prestige. Medieval jurists are moreover very attentive to the goings-on outside their studies and lecture halls, and acutely conscious that they bear the weighty yet honourable burden of bringing order to the potential chaos of the medieval socio-economic sphere.

It is this open attitude to wider society that allows the flowering of medieval legal scholarship to exist in a fertile relationship of symbiosis with the system of customs and facts which continues to underlie medieval law; indeed, scholarship even contrives at times to extend the reach of customary law. The pages of scholarly works provide a stable theoretical and technical home for the novel facts of social and economic life: the new situations encountered in legal practice, and the new legal formulations and the new institutions which lawyers require to cope with those situations. These facts of daily life are constantly forged and reforged in the ever-busy workshop of change that is late medieval society; the work of scholars removes facts from the furnace of change and discovers in them a higher, more rigorous, more universal message. Scholarship makes the legal formulations and institutions of legal practice into models that can be deployed in other similar contexts and at other points in time. In so doing the work of jurists takes upon itself the function of ordering the law, a goal it fulfils completely.

The Character of Late Medieval Legal Scholarship: *Ius Commune*, Roman Law, Canon Law

We must now descend to the level of concrete details. We have seen that legal scholarship plays a central role in the late medieval legal system, and we have seen why this is so. Now we must give a more detailed account of medieval legal scholarship's defining features.

First of all we should note the *isolation* of medieval jurists. By this rather unexpected term, I mean that our community of scholars does not operate within an all-encompassing political sphere like those that we will examine in the modern period: a political system which governs its subjects and imposes conditions upon them, protecting its lawmakers and conferring authority upon them using a police force and its powers of coercion. The authority of jurists was generated entirely by their spiritual and intellectual prestige, and yet how did they maintain the necessary level of observance of the law in the wider community? No one jurist, whether an exalted courtier to a prince or an influential member of the public bodies of a city-state, could satisfy this desperate need for authority. Authority could only be achieved by the plurality of voices contributing to a scholarly consensus.

The jurists still harboured in their unconscious the naturalistic and factualistic convictions of the pragmatic early Middle Ages, which had settled in the collective imaginary. Late medieval jurists thus studied the facts of their contemporary reality closely and with pleasure, because it was in facts that they might detect the primary quality of their work: *effectiveness*. However, they lacked their own structures of *validity* – that is a higher, general and authoritative model into which to incorporate the multitude of pragmatic conceptual and technical solutions for which they had had to reach for the sake of their work's effectiveness. This model was provided by Roman law.

During the early Middle Ages, Roman law had scraped by, donning the ragged clothes which befitted its forgotten status. Roman law became 'vulgarized', as Romanists call it, absorbing the simple, factual, effective traits of its social context and letting the high pinnacles of refined legal erudition fall into disrepair. To give an example: the most precious resource of Roman law, the fifty volumes of Justinian's *Pandecta*, which held the treasures of classical legal scholarship, were unknown throughout the early Middle Ages. They were incomprehensible because they were of no use: farmers and shepherds have no need of feathered hats and sequins. Legal historians rightly emphasize the year 1076, when a Tuscan legal document refers to the *Pandecta* for the first time since antiquity. The return to the *Pandecta* indicates that it was once again comprehensible: it could be put to use with understanding.

The cultural wealth of the eleventh century has a specific consequence for the history of law: the cultural inheritance of Rome that seemed to have been lost forever resurfaces with the plenitude of its scholarly detail intact.

The schoolmen wasted no time in grasping this opportunity. What we have up till now been calling, somewhat vaguely, 'Roman law' is in fact a system of laws codified by Justinian in the first half of the sixth century after Christ in the form of the majestic work called the *Corpus iuris civilis* ('Body of Civil Law'), made up of the *Institutiones* ('Elements'), the *Pandecta* or *Digesta* ('Pandects' or 'Digest'), the *Codex Justinianus* ('Code of Justinian') and the *Novellae constitutiones* ('Novels'). The jurists of the eleventh and twelfth centuries did not regard these texts as a mere treasury of useful terms, rigorous concepts and technical solutions, grounded in a robust and consistent legal language; rather, the *Corpus iuris* was their longed-for authoritative model, to which they deferred.

The *Corpus iuris*'s authority was great for two reasons. Firstly because of its age: it belonged to and contained within its pages the fabled world of the ancients. Antiquity was held worthy of especial veneration: as the above-mentioned John of Salisbury says, *venerabilior est vetustas* ('age should be respected more'). Secondly, the legislator in question was Justinian, who was not only a Roman emperor but moreover an unimpeachably Catholic one. Justinian cherished orthodoxy and protected, or even over-protected, the Church – as can be seen in the learned and respectful manner adopted by those parts of the *Corpus iuris* which deal with Church dogma.

Roman law served as an excellent means by which to justify the validity of medieval legal scholarship: it was the authoritative model which would, in turn, guarantee its emulators' authority and therefore widespread compliance with their work. The scholar had to present himself as a student of classical culture and drape his assertions with the protective mantle of some fragment of the *Corpus iuris*. These fragments were even known in the Middle Ages as *leges* ('laws'), as if to underline the necessity of conforming to them.

So far so good, but what happened to the facts of contemporary existence under these learned scholars? Was the veneration of Roman law not a betrayal of the primacy of effectiveness which these well-integrated jurists still supported enthusiastically? Let us not forget that the Justinian legal code dates back to the sixth century AD and, despite its great scholarly worth, was rooted in a very different society and historical environment from that of the late Middle Ages.

And so the jurists of late medieval Europe were able in good faith to swear fealty to two lieges: the venerable Justinian Code and the demands of contemporary society. This opened up the possibility, of course, that the demands of daily life in the Middle Ages found no answer in these sacred texts or that the answer they did find was unsuitable. The solution in this case highlights the problematic nature of their approach and may be somewhat difficult for the educated twenty-first-century reader to accept: medieval jurists were careful never to depart from the *form* of the Roman *lex* in question, but they often departed from its *substance* where they found it necessary to do so in their role as constructors of a new legal order.

The jurists read the *Corpus iuris* with the eyes of late medieval man; they interpreted it in the light of the novel demands which pressed upon them. In effect, their interest in classical culture leant more towards style than substance. As we shall see, it is this anachronism that will later lead the humanists to scorn the scholastics as ignorant and asinine, since their interpretations of the classical material were not faithful to its original significance and indeed were often a travesty of it. Certainly medieval jurists read Roman law in a deeply contradictory way; only by accepting contradictions could the two sources of law, contemporary facts and the ancient *leges*, coexist. But the jurists still could not and did not discard Roman law and leave it to moulder. This was partly because of the veneration with which they viewed it, but mainly because it granted validity and coherence to their doctrinal musings.

And so medieval jurists found themselves in a curious position: whilst paying enforced and constant homage to the form of Roman law, where necessary they would propose audacious solutions demanded by the present circumstances. Effectively, they would invoke these ancient texts in the service of legal arguments that were at best foreign to their substance and at worst entirely contradictory. Our jurists were certainly *interpreters* of Roman law, but not in the modern European sense of that word, denoting readers of a text who allow the text itself to condition their responses. Their *interpretatio*, and I use the Latin term here for clarity, is more of a mediation between ancient law and novel facts than an explanation or exegesis of the source texts.

By tradition the first great medieval school, established at Bologna in the late eleventh and early twelfth centuries, produced *glossators*, whilst the more mature, culturally rounded institutions set up in thirteenth- and fourteenth-century Italy and Europe produced *commentators*. In either case it should be clear that, whether we are dealing with glosses or comments, the schoolmen of these institutions consistently styled themselves as *interpreters* of the Justinian Code. However, their mediation of Justinian's work was creative: it forged a new law, which typified a historical moment rich in scholarship. We call this law *ius commune* (literally the 'common law', although this is not the same as the English tradition of common law, to be dealt with in the next chapter).

Ius commune was a law created by jurists, by those steeped in legal learning – judges, notaries, advocates and above all scholars. These were schoolmen who taught at universities across Europe but who were fully immersed in the tangible nature of the legal experience. They did not hesitate to make themselves available, whether as advisers to those who wielded power; as legal counsel to the parties in a case or to the judge; or as practising advocates or notaries. The *ius commune* was born out of the complex dialogue that these jurists set up between the facts of contemporary life and the rules laid down in the texts of ancient Rome. It would be fascinating to enter into the precise innovations of the various structures of the *ius commune*, but this book aims only to sketch a general picture of the historical developments. We shall limit ourselves to describing the degree of creativity and imagination shown by the jurists in one very important social and economic field: the law of property. Despite the clarity of Roman sources on the indivisibility of *dominium* ('ownership'), the late medieval readers of Roman law were also the heirs of the early medieval practices which had shifted the emphasis away from the principle of ownership and towards its effects. Late medieval jurists were conscious of the need to come up with a formal legal justification of the present situation, and so they confidently seized upon certain Roman texts and managed to twist their message so much that they were able to build two different forms of property rights out of the same concept of *dominium*. Situations of effective use of goods were now elevated to the rank of *dominium utile* ('ownership through use'). This gave rise to the long-lived theory of *divisible property*, which survived up to the eve of the French Revolution.

The *ius commune* was a pluralistic endeavour which spoke with the voice of an entire community of jurists and knew no borders. This late medieval law without a state can be likened to the handiwork of a class of skilled tradesmen engaged in the construction of a large building. The great Italian legal historian Francesco Calasso has rightly talked of 'the *ius commune* as a spiritual fact'.[7]

The *ius commune* was, as we have said, a law without borders, as is proper for a scholarly discipline. It always searched for universal solutions and rejected artificial political barriers, as the extraordinary circulation of teachers and students in late medieval Europe demonstrates. These cultural pilgrims travelled from one university centre to another, and claimed citizenship of a republic of letters to which all mankind might belong. The *ius commune* set up a universal framework of laws that claimed sole legitimacy through scholarship and effectively unified the legal system of Europe. To give one illustrative example amongst many, let us turn to the *Commentarii* ('Commentaries') of Bartolo da Sassoferrato – an Italian jurist of the early fourteenth century and head of the school of commentators. The *Commentarii* are predominantly made up of lecture notes which bear witness to a lively dialogue between Bartolo and his students. At one point, Bartolo substantiates one of his explanations by reference to a German scholar who, he claims, had mentioned the opinion of a professor at the University of Orléans in his own lecture that morning.[8] To the jurist, the little classroom at the University of Perugia, where Bartolo teaches, is not encircled by the walls of a central Italian city but is, rather, at the centre of a web of intellectual relationships located in space across modern-day Italy, Germany and France – in effect the centre of the entire civilized world.

There are two things we should make clear.

Firstly, that the *ius commune* was not, as one Italian Romanist once put it, 'an updated Roman law', or, as some would reductively argue, the *ius romanum medii aevi* (the 'medieval continuation of Roman law'). Roman law was certainly a source of authority for the *ius commune*, as well as an indispensable guarantor of validity, a necessary point of comparison and a linguistic, technical and conceptual model, but it would be a mischaracterization of medieval law to identify it completely with its Roman predecessor. With regard to the new legal tapestry of the late Middle Ages, Roman law is only one thread in a larger whole, albeit an important one. And as anyone knows, the individual threads that make up a tapestry are transformed into a new and different artefact when they are woven together.

Secondly, and no less importantly, the *ius commune* is so called because it belongs to all the people, and is founded in the rationality that is scholarship's greatest weapon and resource. But it is the *ius commune* also because the medieval concept of personhood combines the citizen and the believer to make a political subject who is equally legitimately governed by the laws of the hegemonic religion – the canon law. A modern jurist would regard this as a set of laws belonging to an organization distinct from the state, with its own independent existence from any state or group of states. But in the Middle Ages such a distinction is unthinkable, since the law of the Roman Church complements that of the former empire, providing a second authoritative model and second pillar of validity. The jurist is thus the interpreter not only of the *Corpus iuris civilis* but equally of the canon law, in which any competent legal scholar must also be well versed. The resultant law is thus 'common' in a second sense, in that it stems from two traditions.

The Reform of Canon Law and the Creation of Classical Canon Law

We have already seen how, during the first millennium, the Church of Rome slowly but steadily constructed a body of laws that suited its purposes, and how the resulting legal system reflected the Church's pastoral mission. We have also seen how, at the outset of the second millennium, a few canon lawyers of the Gregorian period made a preliminary attempt to harmonize the discordances in the canon law system and to revise it in accordance with the fundamental distinction between divine and human law.

The two centuries that follow constituted a coherent and visible implementation of the Gregorian policy. The popes of these centuries were predominantly trained canon lawyers, and they oversaw a complete review and reform of canon law, giving rise to what is often rightly called *classical canon law*. In the mid twelfth century it was a lone monk, Gratian, writing in a private capacity, who attempted to bring consonance to the dissonances of the canon law tradition in his famous work, the *Concordia discordantium canonum* (the 'Harmony of the Clashing Canons'). In the thirteenth century, meanwhile, it was the popes who took it upon themselves in their official capacity to promote significant collections of laws: Gregory IX in 1234, Boniface VIII in 1298 and John XXII in 1317. The Church thus began to gather together what would become known, in an echo of Justinian's great work, as the *Corpus iuris canonici* (the 'Body of Canon Law').

The technical features of this very particular form of law, shaped by the specific demands of the Church's pastoral mission, were those collected very adeptly by Ivo of Chartres, as recounted in the section above. I should now like to add a few considerations on the predominant, although not the sole, type of primary source in which the canon law of this period appears: the *decretal*. *Decretal* is an adjective, meaning decisive, which presupposes the noun *letter* (*epistola decretalis*, 'decisive letter'), and signifies the response given by the Pope to a request for definitive clarification of a doubtful point of canon law that had surfaced in day-to-day life. These strange sources appear difficult to classify to modern eyes, since they are neither laws, nor judgements, nor administrative acts. Instead they typify the combination of powers wielded by the pontiff: they concern a single case upon which they issue a ruling, but they also bind the rulings in similar episodes in the future.

Although the decisions of ecclesiastical councils, the scholarship of canon lawyers and judicial pronouncements all remain important, it is clear that the pontiff now plays a central role. He is the *vicarius Christi* ('vicar of Christ'), the supreme guide and successor to Peter, who can deploy any of the Church's powers – as the massive proliferation of decretals attests. This proliferation also demonstrates that canon law disdains abstract and general rules, preferring instead to focus on the concrete case, with all the garnishings of circumstance that accompany it. The 'pastorality' of canon law, in effect, leads to a law based on casuistry and precedent. From the thirteenth century until the first *Codex iuris canonici* of 1917, the law of the Church of Rome is and will be described here as primarily a *ius decretalium*, a law made up of decretals.

Ius Commune: Special Local and National Laws and Statutes

Its basis in scholarship meant that the *ius commune* was equipped with a startling capacity for expansion. It provided an almost inexhaustible reservoir of technical legal analyses and solutions, of concepts and principles that were suitably abstract and malleable to meet the urgent needs of the late Middle Ages' complex socio-economic reality. The basis of the *ius commune* in reason gave it a universal scope; it was not weakened by local legal customs. Although it was certainly absorbed by very diverse local traditions, every jurist, whether a theorist or a practitioner, was able to draw from it tools and solutions suited to his legal innovations.

The *ius commune* was born in the culturally fertile terrain of north-central Italy, specifically in the University of Bologna: the alma mater of legal scholarship. It then spread out across the whole of Europe, uniting it under one legal vocabulary and set of concepts and so allowing any jurist to feel at home wherever his travels across the politically fractured continent took him. The *ius commune* was taught not only at Bologna and in north-central Italy, but in all the universities of Europe: Salamanca, Lisbon, Montpellier, Orléans and Paris. Fredrick II, the same legislator of dubious quality who bequeathed to the Kingdom of Sicily its own body of law in 1231, stipulated the *ius commune* as the primary object of study in his reform of the law schools of Naples and staffed those schools with teachers trained at Bologna.

But the *ius commune*'s reach was even more pervasive than this: even when drawing up acts of royal legislation, the princely chancelleries, which were full of legal scholars, often based their work on the instructive technical practices of the *ius commune*. The same process occurred during the written drafting of customary laws or city statutes carried out by professional jurist-draftsmen. An even more extreme example is provided by the king of Castile, Alfonso the Wise, whose legislative legacy, *Las siete partidas*, was a text of almost pure *ius commune* translated into the national language. In France also, in the middle of the thirteenth century, the division of the territory into two regions – a south, ruled by *droit écrit* (or *ius commune*), and a *coutumier* north, ruled by predominantly oral customs – was consolidated and remained in place for the duration of the Ancien Régime.

Certain Italian legal historians have recently argued that the *ius commune* was more of a chimera than a true historical presence. I have no hesitation in deeming such scholars to be deluded by an unsustainable fascination with originality, a concept that finds no analogues (and indeed many opposites) in an objective consideration of late medieval society.

Instead I must address the historically salient problem of the relationship between the common law and local legislation, or *ius commune* and *iura propria*. This conflict arises because of the simultaneous presence in the same territory and under the same political system of one type of law that is universal and one that is local. The problem becomes more pressing over the course of the thirteenth century, when the first, admittedly timid, efforts at legislation by kings appear, coupled with a lively flourishing of *statuti* passed by cities, predominantly those of north-central Italy. It is above all in these Italian city-states, rather than in the monarchies, where the friction between common and local law is most keenly felt. For now, monarchs tended to concern themselves with matters of public import ignored by the *ius commune*, or dealt with only scantly. The city-states, meanwhile, had only recently emerged from the sway of empire after a bitter struggle; they drafted statutes with a much wider compass, although still somewhat haphazard and lacking in any aspiration to completeness. These statutes squarely address the common law/local law issue, deciding for the precedence of local law.

Does this mean there was a hierarchy for sources of law? That is what we would have to conclude if we saw the medieval Italian city-state as a sovereign entity when it declares the precedence of its own laws over the *ius commune*. A sovereign state is a rigid monist; it attributes the status of law only to those acts made by itself and tolerates no competing production of law within its borders. Yet I have already shown here how such an interpretative model of the state is unsuitable and misleading in the medieval context, and have instead sought to evoke the medieval legal experience by dwelling on one of its most characteristic features: legal pluralism. Within the same political entity there can be various producers of law, because the politico-legal medieval outlook of the Middle Ages does not provide for political power to be concentrated in the hands of a single officeholder.

In any large *comune* of the thirteenth century, the civic laws, or statutes, were not the only source of legislation: there was also the canon law laid down by the Church; mercantile law set by the community of merchants; and feudal law produced by those of the feudal class. Each of these had its own specific rules governing specific subjects and people and adjudicated by specific tribunals. Finally, there was the *ius commune* – constructed from the interpretation of the 'universal laws' (Roman and canon) by the universal community of jurists. The civic political order was unitary, but within the city walls also dwelt plural, diverse legal orders which coexisted with one another and shared in the government of the city's inhabitants.

The law was not held in the smothering embrace of the apparatus of public power; instead it led society, expressing its desires and conditioning its actions. The reference in the statutes to the *ius commune* does not set up a hierarchy because there can only be such a thing when political and legal orders coincide in an indivisible sovereign power. The *statuto* is a complex of rules inscribed under the banner of the concrete, which services the needs arising in civic life; these documents limit themselves to clarifying that, in any matter, if there is a civic law dealing with the issue, the judge should apply that law, but, if there is a gap in the civic law, the judge should draw on the omnipresent and theoretically complete *ius commune* that needs no authorization to fill it.

In this integrated plurality of legal systems, *ius commune* and *iura propria* are examples of unitary legal orders that are not undermined by their proximity to power. The wielders of power, meanwhile, followed the collective consciousness in recognizing that the unity of the law went beyond their sphere. One great Italian legal historian whom I have already mentioned, Francesco Calasso, talked of a 'system of common law', made up of the *ius commune* and the local legal orders, which were not isolated one from another but rather part of a permanent integrating dialectic.[9] Calasso's interpretation is convincing if we interpret that demanding word *system* to mean something similar to what we have found to be rigorously true in the course of this chapter: a sense that both universal and particular are incorporated in a greater unity that respects plurality and diversity.

Ius Commune and Feudal Law: On the *Usus Feudorum*

The term *medieval civilization* tends to be accompanied by a further adjective, *feudal*, and with some justification. However, I should like to take the opportunity to clarify to the reader the significance of this very particular word. Firstly, it should be noted that the existence of a feudal class is not only a European phenomenon: it has arisen every time a historical civilization has found itself in similar socio-political circumstances, for example in the Chinese and Japanese empires – places that felt no European influence until the modern era.

The origins of what we now call the feudal order can be traced back to the origins of medieval civilization itself. The primary cause of feudalism can be sought in the way in which the political order adapted itself to the nascent medieval historical context. I shall reiterate here what I have already argued so far, with added nuance. The political and legal class of the Middle Ages is characterized by the following features: the impotence of the central authorities and their incapacity to impose their will, and the growing influence of other powers both by their *de facto* occupation of positions of strength and by formal entitlements granted from above. Amongst these other powers, economics stands out: the possessor of wealth has access to the only decisive force in Middle Ages and, in a very slow process, he gradually gains the offices of judge, military commander and tax collector in his own lands. The Middle Ages are truly the historical moment in which the divisions between private and public are most fully erased. Many of those who wielded power from afar were in fact obliged to delegate that power to those more immediately present on the ground. This exacerbated the fracturing of political power in the Middle Ages, with the result that the political order was made up of a complex network of relationships that were only at first glance hierarchical.

In the legal sphere, this hierarchical structure, although belied in effect by the reality on the ground, was communicated formally via relationships of superiority and inferiority. The superiors promised protection and the inferiors swore loyalty via a series of links between individuals that often bore little relation to the effective situation of powers in an area of territory. The status of feudatory, or vassal, meant formally that the individual belonged to another man, but often the so-called inferior was, in effect, able to exercise considerable autonomy of discretion.

Feudalism signifies these complex interrelationships of people bound together by mutual bonds of protection and loyalty. The interrelationships soon became personified by a class of people, all of whom found roles in the intricate and fragmented mechanism of powers which linked the highest prince to the lowest serf. This process separated feudal powers off from the general multitudes of common mortals. It should be stressed that this commodification of the network of relationships was a slow process, but it did finally lead to the absorption of the feudal principles of mutual protection and loyalty into the land. There came about feudal territories which incorporated that mixture of public and private which is the primary feature of a feudal structure, with the result that certain public powers (known as *honores*, 'honours') came with the soil and those who acquired ownership of the land acquired with it the powers.

With regard to the legal sphere, I should briefly highlight the fact that, because of its separation, this complicated but isolated web of people and goods soon brought about an even more complex network of customs. Because these customs were restricted to certain subjects and certain areas, they took on the features of an autonomous body of law which we might call *feudal law*. This autonomy was entrenched by the creation of special tribunals to rule in the disputes regarding people from those lands or the lands themselves.

In the middle of the twelfth century, the sum of customs and judicial rulings, by now rendered extremely complex by the centuries-long process of accumulation, was put in order for the first time by an insightful practitioner of law: a Milanese judge. The collection was called the *Usus feudorum* ('Feudal Customs') or the *Libri feudorum* ('Feudal Books'), and its inclusion as an appendix to one copy of the *Corpus iuris civilis* suggests that its material was now considered worthy of scholarly attention. And so scholars did study feudal law, giving rise to writings that are often of great cultural import; the great doctors of the *ius commune* were often not only Roman lawyers or canon lawyers but also feudal lawyers. There are many examples of such scholars: one could cite Baldo degli Ubaldi, a great Italian commentator of the fourteenth century, often acknowledged as the greatest philosopher amongst the jurists and author of a detailed commentary on the whole of Justinian's *Corpus*, on part of the *Corpus iuris canonici* and on the *Libri feudorum*. Feudal law as a special type of universal law came into close dialogue with the *ius commune* thanks to the legal pluralism of the late Middle Ages, with its legal universe that was, as we have seen, both unified and, at the same time, plural.

The Origins of Commercial Law

In legal parlance, *commercial law* signifies the complex of rules and institutions which governs that speciality. It is a field not of use to the general citizen but rather to those engaged in the mercantile profession. Merchants were a growing economic, social and political force in late medieval Europe, and with this new influence they gained the confidence to construct legal strategies to defend their interests. At its beginnings, commercial law consisted in nothing more than the customs of the mercantile class, whose members governed commercial dealings to their own satisfaction. These customs were born out of everyday practices – the dealings of the local market square made general by the now universal esteem in which the mercantile class was held. The customs were written down for the convenience of the users and became, by the middle of the thirteenth century, proper statutes of commercial law, reflective of the now fully realized power of the mercantile class.

Little by little, during the late Middle Ages, we find many developments: the invention of new commercial instruments – such as negotiable instruments, business associations, insolvency and insurance; the streamlining of old arrangements to fit them for commercial purposes – such as agency and assignment of credit; and the overcoming of old stumbling blocks deriving from a now unjustifiable technical analysis of Roman law – such as contracts for the benefit of third parties.

An organic collection of institutions began to take shape and, alongside it, a complex professional mercantile organization. One very significant advance was the creation of special tribunals; at first these had only a limited field of professional and disciplinary activity, but they soon grew to encompass a proper jurisdiction equipped with its own set of rules. Commercial courts were able to rule on any aspect of commercial activity; they were presided over by unrobed judges, and followed procedures that were specifically designed for speed and efficiency. They would be long-lived and difficult to kill off: Italy's *tribunali di commercio* were only abolished in 1888.

Mercantile law was, without doubt, one of the protagonists of the piecemeal system of late medieval law.

2

The Foundations of the Modern Legal System

The Fourteenth Century through a Legal Historian's Eyes: Socio-economic Conflict and Crises of Values

We are accustomed to regarding the fourteenth century as the culmination of the medieval period, dazzled by the brilliance and conceptual scope we find recorded in accounts of the plastic arts and literature. As a Tuscan myself, I cannot help but point to Dante, Boccaccio and Giotto to substantiate this. But, viewed from another angle, the fourteenth century is in fact a period of great disruption, especially if we look at things on a structural level: that of agriculture, nutrition, demography and health. Such things may seem lowlier concerns than poetry and art, but they are often more determining of the conditions of daily life for the average person.

The fourteenth century saw ruinous wars, famines and epidemics, and the persistent, corrosive presence of hunger. The dark protagonist of the century is undoubtedly the plague, whose destruction lasted many years, reaching its peak in 1347–51. The Black Death devastated all of Europe, decimating the population. The result was a widespread abandonment of lands, a rapid decline in agriculture and an increasing, albeit futile, flight to the cities. At the root of these structural problems is the fact that mere survival became difficult for the average person, which in turn dealt a destabilizing blow to the collective consciousness.

We are not dealing here with a sudden and wrenching shift, as a hasty account of the situation might lead one to believe. The old beast begins to look a little tired and inside it a few cancerous cells are born which it harbours and nourishes. The old carries within it the seed of the new, and thus fosters its own demise. The fourteenth century thus appears to be that which it is, historically speaking: a period of transition, in which old and new intermingle, and the outlines of a future construction can be perceived. The attitude of a man who knows he finds himself at a moment of historic change is perhaps best expressed by the works of Francis Petrarch (1304–74). The great poet is also a man of culture in whom the seeds of the forthcoming humanistic revolution are beginning to take root, as can be seen in the extraordinarily lucid passage in which he avows that he feels he lives in a borderland that forces him to look at once backwards to the dying world and forwards to its nascent replacement.[1]

It is obvious that the collective consciousness began to doubt the load-bearing pillars of the old order, which appeared to be about to buckle. Both nature and culture seemed to have betrayed late medieval man by failing to guarantee his survival. This led to a belief in the need to reform the old social, political and legal order from top to bottom, emphasizing a new set of values. The fourteenth century is a very singular point in time: the response to this crisis in underlying structures takes the form of a theological and philosophical reflection, which attempts to reform completely the anthropology of the Middle Ages. The emphasis is placed squarely on the individual, but this figure can only be trusted if freed from the chains that have bound him for the entirety of the medieval period.

The process which set out from this crossroads crowded with structural changes and visions of intellectual reform was fundamentally a process of liberation. The new anthropology was decidedly liberating in character: whether in the freeing of one's brute nature, in the freeing of individual subjects from their many social bonds and from the social order which had both protected and constrained them, or in the liberty from the world of things which had let down medieval man in their failure to lead him to salvation. The philosophical debates which arose and grew to full volume during the course of the century all have at their heart the idea of isolating the individual subject from the world and in the world: they all recognize the human subject as capable of searching far within himself and finding sufficient strength to assert dominion over reality.

The medieval human subject, as analysed thoroughly and adroitly by Thomas Aquinas, is an *intelligent* being – identified above all by the power of reason. Above all, he possesses *understanding*, a quality which allows one to project one's thoughts beyond oneself, in a psychological gesture of humility towards the surrounding world. The new human subject, who will be most fully defined by Franciscan theology and philosophy, is seen as a being who *loves* and who *wills*, a subject whose identity is predicated on the most autonomous and self-referential aspects of psychology, so that autonomy defines identity. Everything becomes subjectivized and resolved within the boundaries of the subject, who thereby affirms his ontological separation from the world and consequent liberty within the world.

It is instructive to examine how this rediscovery of liberty is consolidated: by the self-determination of the will, which is conceived of in legal terms as a *dominium*. It is certainly undeniable that in classical Roman law, the concept of *dominium* is not merely political but strongly linked to the freedom of the individual subject, in contrast to the openly economico-legal analysis of the term in the medieval period. The new anthropology that develops during the heady fourteenth century, and afterwards, is predicated on the universal applicability of the concept of *dominium*. This becomes the general analytical category into which both intersubjective and intrasubjective realities are classified. Alongside the ownership of things in the exterior world (*dominium rerum*), we can observe a much greater emphasis on the idea of *dominium sui*, the right of ownership that every person has over his own body and skills. This right of property stems from the divine duty instilled in every human creature to protect its own existence. This right is characterized by a considerable capacity for expansion since the being that possesses it is entitled to assert itself over the cosmos and the surrounding culture. The will, in effect, becomes the essential character of the human subject, and the guarantor of each person's liberty. This liberty is construed as *dominium*; the intrasubjective realm is governed by a series of rights of property.

If the reader is feeling impatient with the apparently generic arguments put forward so far, I would reiterate that it is in this nexus of fourteenth-century anthropological developments, which stand in polemical contrast to various medieval certainties, that we can grasp the origin of the individualism which will dominate modernity absolutely. These anthropological considerations allow us to capture the first instance of that typifying characteristic of modernity: the mingling of being and having, of me and mine. The act of possession is hereafter seen as contributing to the very being of the human subject. We shall deal with these developments in greater depth below.

Another probable objection to what I have argued so far, and it is a legitimate one, is that I have dealt with historical phenomena in the areas of agriculture, demographics and health which contributed to the fourteenth century's structural crisis, and with abstract philosophical choices contributing to an anthropological shift, but I have not yet dealt with the law, or even with its close neighbour, politics. To borrow from Hegel a figure of speech that I have always thought very felicitous and apply it to our field, I would reply that the law is like the owl, the bird of Athena, because it does not seek out the midday sun but waits until the heated business of the day is over to spread its wings. The law, as I have argued many times, is a reality of the most rooted kind: it lies in the most profound layers of civilization, where the very values of society are found. The law emerges into the sun only when problems on the social, economic and political surface become revolutions that reach down to the deepest, most radical level.

The period we are examining is a borderland between two experiences of the law; let us therefore reiterate here two considerations that are relevant to our understanding of the historical phenomena. Firstly, an *experience of the law* means a typical way of conceiving of, feeling and living the law – it is not limited or limitable to the laws passed by a monarch, or the commands made by members of a police force; it is, rather, a cultural reality. The medieval experience of the law is not an inconsistent but harmless period of transition: it lasted for many centuries and penetrated the consciousness of medieval man. The medieval experience of the law slowly succeeded in changing people's values by inculcating its understanding of anthropology. The structural crisis of the fourteenth century was a historically propitious moment, which favoured a recasting of the collective consciousness, a renewal of mankind's vision of its place in the cosmos and in society. This renewal could not but begin with anthropology: the *reicentric* anthropology of the Middle Ages had to be replaced by one that was *anthropocentric*, and this is what the voluntaristic currents in philosophy and theology began to propound in the fourteenth century. The law followed later, in perfect harmony, once it had absorbed and assimilated the results of these currents.

Now we are beginning to examine European legal modernity, I should warn the reader that the quest to found a new legal order lasted some centuries; the slow but progressive construction of a fully realized model ended in a completely renewed legal order only at the end of the eighteenth century, when the reforming winds of the French Revolution cleared away the detritus of the old era from the streets of Paris and continental Europe. For reasons of inertia, that socio-political order which we call by its French name, Ancien Régime – that is the French state up to 1789 – retained many relics of medieval law, which intermingled with the increasing numbers of juridical innovations. A completely new legal order, however, only appears after the great revolution.

A Process of Liberation: The Rise of Macro-individuals, Micro-individuals. The rise of the Nation-State

The theological and philosophical emphasis on the will which began to circulate in the more unfettered culture of the fourteenth century had a great effect at the level of the collective consciousness, becoming a very real current in political thought. This voluntarism struck the first blow for liberty, sweeping away the network of medieval relationships between individuals which prevented each from asserting his independence: a process which altered the subject in every field – religious, economic and political.

This new process of liberation, which began in the fourteenth century, aimed above all to liberate the individual. The defining characteristic of modernity can be seen in the desire to set up a new political system based not in the medieval preoccupations of nature and community – and the communitarian solutions that went with them – but rather in a vision of society as a collection of human individuals, whose freedom is recognized and respected. These individuals were free to dominate the realities of nature and of society, each of which gained a strongly felt psychological force.

Above all, the new society focused on the physical subject, who was now psychologically liberated and no longer needed to cower under the covering of protective communitarian structures, which therefore began to feel smothering. Until the end of the Ancien Régime, society was in fact an aggregation of societies, a complex melting-pot of intermediary structures. Society was often seen, therefore, as an impediment to the freedom of both the private micro-subject and the public macro-subject, and indeed it had effectively become an impediment to the free action of the self upon the other.

The physical individual, now psychologically affirmed, soon made robust gains in the fields of culture, religion, economics and law – as we shall see. However, since the individual's liberty was founded on the dominance of the will and identified in the proprietary rights one had over one's person and external goods, the economic sphere became especially important. The effect of this is that the defining figure of modernity is the property owner.

The public macro-subject just mentioned was a political being smothered on all sides by society and by social institutions that were an encumbrance, because they had been founded in a different age, but which were nonetheless difficult for a prince to abolish. Towards the beginning of this book, I emphasized that medieval political power was incomplete, and refused to use the term *state* in the medieval context in order to avoid confusion when dealing with a type of political power that had to come to terms with the enormous and overarching power of medieval society.

Now that we turn to the fourteenth century, we find that the public macro-subject has a few more weapons at its disposal. Amongst the institutions that came under attack are those universal networks of the Middle Ages, the Church of Rome and the Holy Roman Empire – the former of great import given its monopoly on sacred matters, the latter now reduced to a vestigial state but still capable of exerting influence if someone enterprising took the imperial throne. The fourteenth-century political sphere became completely overwhelmed by the process of the liberation of the individual, and the ragged remains of the previous political fabric were knit together into individual political entities possessed of an unprecedented degree of will. These are not yet states or, if they are, they are only so in embryonic form. But this era certainly marks the beginning of the modern state; it is at the start of a journey which, from here on, develops towards the statism of the seventeenth and eighteenth centuries. The process can be described simply: the more the state grew, the more society shrank.

Princes and the Law: The Kingdom of France in Detail, a Testing Ground for Modern Law and Politics

The kingdom of France stands out amongst these political entities that affirmed their own individuality and threw off the heavy trappings of medieval universalism. France is a true testing ground for modern law and politics; here what will one day become a widespread movement sees its first pioneering manifestations.

I have already highlighted how, during the thirteenth century, the enlargement of the territory of France from the reign of Philip II Augustus (1180–1223) on, led to a desire for total independence. The jurists of the royal court, meanwhile, affirmed the absolute sovereignty of the monarch over temporal things. In the mid thirteenth century, St Louis, an adroit king, who was conscious of the possibilities of his position, had no hesitation in deploying an eloquent official language in order to define his power as totalizing: *plénité de la royal puissance* (the 'plenitude of royal power').[2] However, the former legislative timidity continues, since the king is still hindered by the strength of the feudal classes

It is at the turn of the thirteenth and fourteenth centuries, during the reign of Philip the Fair (1285–1314), that the French king was able to rid himself of all encumbrances and checks on his power. Philip launched a bold programme that was at the limits of audacity, as is demonstrated by the conflict that it provoked with the pope of the time, Boniface VIII – a pontiff who denied the historical forces at work in the early fourteenth century and insisted on claiming the primacy of papal authority in both spiritual and temporal matters. In 1302, when Boniface had solemnly reaffirmed his anti-historical and theocratic intentions in the bull entitled *Unam Sanctam*, Philip the Fair brought together a very broad convocation of prelates, barons and delegates from cities who united in recognition of his sovereignty. It is by now appropriate to apply that loaded term to the status that Philip claimed: *sovereignty* denoted the absolute supremacy of the king in the temporal sphere; his lay and ecclesiastical vassals swore fealty directly to him, and he was granted authority over his realm directly by God. The year after that, Boniface died and the king soon gained the upper hand with his successors – especially after 1305, when a French former prelate ascended to be Clement V.

The fundamental point to recognize here is the emergence and consolidation of an authentically sovereign power, which presaged the future states of modern Europe. From a legal historical point of view, the first thing to highlight is that this power appears ever more complete, that is to say it shows a tendency to expand and to bring under its auspices an ever greater social territory. The character of this supreme power has changed noticeably: the *ordonnances* have multiplied and the king, as dispenser of justice, now attaches great importance to the production of laws, and places it amongst the most important aspects of his prerogatives. This was the beginning of a historical current lasting several centuries that would only reach its conclusion during the reign of Napoleon I: the epitome of the sovereign legislator, a codifier of laws who succeeded in bringing all laws, even private ones, together in a single code.

Let me be clear: fourteenth-century French monarchs were certainly concerned with the law – they perceived its precious value as a means of cementing their power – but they still acted with caution, which we can see because their legal documents still base themselves in the material facts of their realm and leave the governance of the private relationships of their subjects to immemorial customs. Nonetheless, from the reign of Philip the Fair onwards, the mindset of political power in France changed: the qualities of *aequitas* that were fundamental to the medieval vision of the prince faded in importance, and the first amongst his tasks became 'the making of new laws or constitutions for his subjects', according to the unknown author of the *Songe du Vergier*, a late fourteenth-century politico-legal compilation which argued that the French king had equivalent legislative powers to the Roman emperor.[3]

This new mindset brought with it a suspicion regarding the common law, which was by definition a scholarly and universalizing law, despite the royal jurists' frequent echoes of the absolutist rhetoric of the emperors contained in the *Corpus iuris*. In 1312, when Philip the Fair reformed the study of law at the University of Orléans, an institution already famous for its illustrious teachers, he was careful to specify that the *Corpus iuris* possessed authority within his realm only because it constituted a customary form of law allowed its force by the king.

Yet it is in the specific field of *coutumes* ('customs') that the range of the new power's influence can be seen. The customs were not suppressed, or even modified – as I have said, the kings proceeded with caution – but there is a more indirect assertion of royal influence by means of a comprehensive written drafting of the sprawling mass of oral customs. The process was somewhat submerged at first and was made official only in the mid-fifteenth century: in 1454, Charles VII ordered a full written drafting of the oral customs. The resulting procedure consisted in a continual intervention by the central power. Formally the 'constitutional' basis of the customs remained untouched, but the monarchy achieved two goals that were of great importance in French legal history: they asserted a tight control over the customs and they began to bring some unity to the highly fragmented field of customary law.

Francis I's ordinance of Viller-Cotterêts in 1539 reformed the legal system by ordering that all official legal acts, including compilations of *coutumes*, had to be written 'in the French mother tongue and no other manner'. A specifically 'French' law now begins to appear, a linguistically autonomous, concrete entity that reflects the unity and independence of the French state.

This should not be taken to mean that the monarchy reduced the law to the single dimension of legislation. Although the quantity of royal interventions in the legal sphere certainly grew, the legal pluralism of the Middle Ages remained and was eradicated only by the French Revolution. In 1576 the jurist Jean Bodin, in his justly celebrated *Six Books on the State*, an acute and lucid diagnosis of the new legal and political structures, records a complex legal landscape which he divides into two fields, *loy* and *droit*: the former describes the express will of the monarch as manifested in authoritative commands; the latter denotes the rules handed down by tradition, born of things and infused with the underlying equity of nature.[4] At the end of the sixteenth century, the historical current referred to earlier is certainly under way.

The Kingdom of England in Detail: The Origins of the English Common Law and its Continuity with the Medieval Outlook

England is a nation located on an island, separated from the European mainland by a deep stretch of water. This may appear an absurdly obvious statement to make in geographical terms, but, when transported to the legal historical sphere, it loses that absurdity. The insularity of the English historical context with regard to the European *mélange* is notable. I shall seek to give an account, which will be brief because of the demands of the present work, of how this insularity came about and was consolidated, although it is, of course, a far from easy task to reduce a historical process of such complexity and articulation to a simple linear narrative.

There is one established date and historical fact that can serve as a point of departure: the defeat in the Battle of Hastings of the last Saxon monarch by the Norman invader, William the Conqueror. The Norman Conquest is an event which is traditionally identified as the moment which transformed the political and legal landscape of English society.

Before the Conquest, and despite the odd distinguished monarch, English political history was turbulent because the land was beset by frequent and violent intrusions by Viking invaders. The legal regime of the time consisted of a scattering of provisions based in custom. After the arrival of the Normans, however, a unitary political structure was set up to head a somewhat incongruous central power base.

What is interesting from our perspective is that this centralism is not limited to the political sphere, but also affects the law directly. Centralized royal courts located in London are set up which immediately blend in to the wider activities of government, all under the auspices of the king. What will come to be called the *common law* (not to be confused with the very separate continental tradition of *ius commune*, dealt with in the previous chapter) is the law common to all free men of the realm. It can have this status because it comes from king, insofar as his power is manifested through the law courts. Although William had sworn at his coronation to safeguard the observance of the existing laws, the new common law gradually replaced the old mass of customs on which local courts would draw. It achieved this not by the formal abrogation of the local customs, but because the greater efficiency of the royal courts which applied the new discipline meant that they became the preferred forums for obtaining justice, rather than their local counterparts.

The resulting legal environment, which consisted in an almost total fusion between government and judiciary, was characterized by frequent royal interventions in legal matters. The typical form taken by such an intervention was that defining innovation of the early common law period, the *writ* – a document sent by the king to the presiding officer of a local court, which required the satisfaction of a specific demand. The parties in a case needed to acquire a writ that covered the specific claim they were pursuing.

There are a few considerations that should be added to this portrait of the common law.

The first is that the common law is a legal system based rigorously on precedent and case law, and characterized by a series of specific legal procedures aimed at remedying specific substantive situations. The most important field in common law is the procedural: the existence of an effective procedural remedy grants the facts of a case their force in law.

The second consideration is that, supporting the holders of offices of political power, there were necessarily legal experts with the technical expertise to craft the writs themselves. It is no surprise therefore that during the twelfth century, particularly during the long reign of Henry II (1154–89), a class of lawyers of sufficient technical expertise began to form. It is very important to point out that, unlike their continental colleagues, these lawyers were not alumni of the prestigious universities. Their knowledge of the law was therefore not an abstract one, based on the Roman model; rather these were legal practitioners who had completed their studies at the Inns of Court in London, foundations which were and are exclusively professional in nature.

Thirdly, because of the lawyers' pragmatic focus and because of the technical understanding needed to craft the Crown's legislation, this growing legal class became the technical means by which the developments in the common law were implemented – the indispensable mediators between the power of the Crown and its subjects. The legal class was increasingly able to remove the more unpleasantly authoritarian, potestative aspects of the royal commands, as if they were digested by the lawyers, who transformed them into a harmonious collection of legal institutes. Over time, these institutes formed a sort of customary law applying to the whole kingdom.

This, by no means simple, legal context also had to deal with the political evolution as represented by the English monarchy. Centralization fostered intolerance, resistance and internecine strife; these led inevitably to some loosening of the Norman monarchy's grip on power. Magna Carta, signed in 1215, is an example of such a loosening; the document has acquired a somewhat pseudo-historical renown, but, with certain reservations, one can argue that at least some parts of Magna Carta are a precursor to the later bills of rights that we shall examine below (see p. 71). And so, during the long reign of Edward III (1216–72), the two chambers of Parliament – the Lords representing the barons and the Commons representing the shires and the common folk – began to acquire greater powers.

These developments are all part of political and institutional history, but they have visible repercussions in the field of law: the power of the judiciary becomes separated from the person of the king and the three courts of the common law become separate from the king's government. The common law is singular in being a law whose authority originates from the king but which, from the second half of the thirteenth century onwards, is distinct from the king himself. Thus comes about what English-speakers now call the *rule of law*, implying the autonomy of the law but also the supremacy of the law over the niceties of politics and the subordination of the state itself to the law. By now the common law is administrated by a class of jurists, advocates and judges, who have their own identity fostered in the cloistered communities of the Inns of Court, but who remain nonetheless rigorously open to the influx of new recruits.

From what has been said above, the reader will be able to see why in the very title of the previous section I dubbed the kingdom of France the testing ground of modern law and politics. The legal history of the kingdom of England, as the title of the present section makes clear, takes the opposite route. In England, the weakness of the central power means, amongst various ramifications, that political power and the law become riven from one another; there is therefore less of a tendency to produce law through legislation than in continental Europe.

Common law, in this instance, means law produced largely by jurists, by those who know the law well and who, in medieval England, tended to be practitioners, men trained in the schools of their own profession. In England, medieval law, that is to say the production of law by a body other than the state, persisted into the modern period. The political sphere certainly did promulgate its own statutes, but these were of marginal import until the nineteenth century. Indeed it was not until the Labour government of Clement Attlee in 1945, which instituted a welfare state and with it the inevitable planning of economic and social relationships by central government, that we can truly see large-scale intervention by the political sphere in England.

Let us attempt to draw a few conclusions from these considerations. We have seen above how medieval law survives in England well beyond the historical boundaries of the Middle Ages. Indeed we may add that it survives the whole of the modern period. The history of the common law runs unbroken from the Norman Conquest to the present day. The suffocating and perverse idolatry of the law and of the legal sphere that afflicts continental law in the eighteenth and nineteenth centuries passes English law by entirely. English law is insular because it is self-generating: it is a practical law made by practitioners and hence estranged from the great cultural movements that flow freely through the law schools and courts of the continent. All attempts by Roman law to 'invade' the 'island' of English law are rebuffed decisively before they can take root, although there are some hints to be found in the work of legal scholars such as Glanvill (twelfth century) and Bracton (thirteenth century). The common law tradition knits together the community of lawyers, who base their sense of professional pride upon it, and draw on it to resist attempts at infiltration from the outside.

The law is a steadfast continuity that underlies all political changes. It is at the root of the English nation's identity and is often inseparable from that identity; it is the most characteristic symbol of Englishness.

An Ideological Break with the Past: Ploughing the Furrow of Individual Freedom. Humanism, the Reformation, Proto-capitalism and the Scientific Revolution

We have seen how the fourteenth century was an era of transition, where medieval civilization started to pass away but new and old coexisted for a while, as the liberation of many types of individual gradually eroded the old order. In the two centuries that follow freedom has already been achieved. During this time Europe stands at the crossroads of great movements of thought and action which consolidate the liberation of the individual at every level.

These are matters whose importance far exceeds the boundaries of the legal sphere, but I cannot leave them unmentioned here if the reader is to have the information that is vital to an understanding of the radical shift from which the modern legal system draws its shape and its substance. I shall of course limit myself here to outlining only the barest essentials of a very complex religious, cultural and economic context that seem to me indispensable to any account of the foundations of the new experience of the law.

1. The first result of the destructive cracks that appeared in the old order in the fourteenth century was the institution of a purely humanistic view of the man–society–nature relationship. Above all, humanism represented a new anthropological perspective predicated on an attitude of complete trust in the idea of the individual subject as the sovereign element in both society and nature. Thanks to the individual's will, ownership of his person and miraculous powers, he was able to bend the world suit his own devices. The defining characteristics of humanism are an optimistic evaluation of the capacities of the subject and a consequent attempt to free individuals from any external conditioning factors, whether phenomenal or social. Humanism therefore fostered a mindset that proudly celebrated self-sufficiency, and sought to eradicate once and for all the detested humility of medieval man. In the popular imagination, humanism is usually associated with the rediscovery of classical culture, and this is not wrong, but it is a secondary feature of the movement; the reverence for the classics highlights the extent to which the *homines novi* ('new men') of the fourteenth and fifteenth centuries identified with the writings of that culture. Humanism itself, however, is essentially a novel interpretation of the world put forward by a now rigorously anthropocentric culture.

2. The broader aspects of humanism came in contact with the religious debates of the era, giving rise to that most emblematic of sixteenth-century historical phenomena: Protestantism and the religious Reformation that follows in its wake. Courageous works by various writers over the previous

two centuries had gradually anticipated and even paved the way for the outbreak of protests that marked the second decade of the sixteenth century. Protestantism soon became the religious arm of humanism, although later on Luther himself would take issue with Erasmus and with humanistic positions, and several other leaders of the Reformation also had second thoughts. Indeed the very concept of *Reformation* is historically unsatisfactory, since the waters of reform coursed down many different channels. Nonetheless, at the root of all the different manifestations of this great religious movement there was a unifying humanistic message. This can be summed up as a demand to free the individual believer from the smothering embrace of the hierarchies prevalent in the religious sectors of society. Meanwhile, the individual believer was optimistically held to be capable of direct and immediate dialogue with the divinity, and also of reading the Holy Bible unaided and without intrusive mediators. In the next section, I shall try to tease out some legal implications from these developments, but for now let it suffice to emphasize the absolute 'modernity' of these changes and the way in which they undermined the communitarian convictions of medieval religiosity. Even in a religious context, the individuality of every subject is affirmed.

3. The far-reaching effects of the discovery of the American continent at the end of the fifteenth century begin to make themselves felt in the early sixteenth century. This epochal event affected even the economic sphere because of the greater amounts of precious metals which begin to circulate in Europe and the broadening and intensifying of trade both by land and sea, both of which wrought extensive changes upon the European economic system. The small-scale dealings of the immediate post-medieval period began to be abandoned in favour of structures and transactions that were much larger in scope. With this shift comes the emergence of a novel and profoundly different psychological outlook: the *homo economicus*.

This historical moment can fairly be called proto-capitalist, where capitalist can be taken to mean an economic system geared towards maximizing profit at any cost. In the sixteenth century, profit has been sufficiently ennobled that it becomes permissible for it to represent the sole aim of a personal endeavour. The medieval merchant may well have accumulated riches, but he was terrified of finding himself in a state of sin, or gambling with his eternal soul. The results of the bad or troubled consciences of medieval merchants can be seen in the many monasteries founded and churches built using funds such individuals gave in their old age or left in their wills in the hope of securing the prayers of numerous churchmen for their souls.

The new economic agent became immersed in the newfound intensity of sixteenth-century commerce. The recent philosophical innovations which located his essential characteristics in the will, in the freedom of the individual and in self-ownership allowed the humanistic businessman to become an emancipated psychological entity, free of the ecclesiastically inspired scruples that had tormented the medieval *mercator*. The sixteenth-century merchant was able to dedicate himself completely and with a clear conscience to the accumulation of riches. Indeed, before long, early modern businessmen would discover that some Reformation theologians were prepared to permit lending at interest (something forbidden by their Catholic predecessors); some even identified accumulated riches as a sign of divine protection.

4. There is a further point to be made, which may at first glance seem unrelated to the topic under discussion, but whose relevance will become clear as my argument develops. There is one other decisive emancipation to bring to the reader's attention, one which develops particularly from the late sixteenth century onward: the natural sciences. Early modern scientists begin to search without any moral or religious inhibitions for the truths inscribed in the complexities of the cosmos that can be deciphered by human ingenuity. The model for these figures is Galileo (1564–1642), who strives to decode the mathematical architecture of the universe, believing that, although its laws may be hidden, they may nonetheless be understood by a person with suitably acute vision, who is able to remove the distorting spectacles of ignorance. I mention Galileo here because of the particular weight his name carries, but there are many other men one could cite.

I should make two further observations by way of conclusion.

Once the natural sciences had developed a consciousness of their own disciplinary autonomy and validity, and particularly after notable successes had been achieved in the field, there soon came about a conflation of the figure of the researcher into higher philosophical and moral principles with that of the mathematician or the geometer who is able to discern the rigorous principles that underpin the physical world. The archetype of this typically early modern form of intellectual is perhaps the Frenchman René Descartes (1596–1650), who uses the absolute rigour of mathematics to perform a philosophical exploration of the human soul. Perhaps even more significant than he is another Frenchman, Blaise Pascal (1623–62) – the subtle and troubled moralist who also wrote influential works of mathematics and physics.

As the fame that now attaches to the name of Descartes may suggest, there took place, during the sixteenth and seventeenth centuries, a slow cultural revolution whereby the exalted place of legal scholarship in instilling methodological rigour in a young scholar was supplanted by the physical and mathematical sciences. During the late Middle Ages, epistemological primacy was granted to legal scholarship since it was the only discipline able to draw on the wisdom of a tradition that was nearly two millennia old and the only one to possess a fully rigorous vocabulary and conceptual framework. So rigorous, in fact, was the study of law that it provided the foundations for the construction of an extremely solid legal system. Of course, to early modern man, medieval jurisprudence was tainted by its association with the medieval mindset, particularly by features such as the fusion of canon and civil law. Medieval legal scholarship therefore suffered the same fate as all the other pillars of medieval culture: it no longer appeared trustworthy, unlike the penetrating insights of the astronomers and the mathematicians, who revealed the secrets of the cosmos to early modern man. Thereafter, even jurists and moral philosophers would follow the methodological example of the natural scientists.

Religious Reform and the Legal System

On 10 December 1520, a fire was lit in the parvis of Wittenberg Cathedral, Saxony. It was a bonfire of books, or bibliocaust, ordered by the first of the reformers, Martin Luther, who was seeking to reduce to ashes the ideas of the books whose matter he had had set alight. Two items on that bonfire are of particular interest to the legal historian: a great collection of tomes that comprised the *Corpus iuris canonici*, and a slender volume called the *Summa angelica*. Let me now explain why those items are of interest.

By burning the pages of the *Corpus iuris canonici*, the legal handbook of the medieval Church, Luther sought to condemn and to extirpate the Church's claims to legal authority, which he saw as implying a claim to political power. In Luther's opinion, this politicking was tantamount to wallowing in the filth of temporality, and he hoped to deny the seal of holiness to the hierarchical structure it had produced. The compendious *Corpus iuris canonici* testified to this claim to legal authority, which represented, in the great reformer's eyes, the betrayal of the Roman Church and the proof of what he called its *captivitas babylonica* (its 'Babylonian captivity', i.e. its enslavement to the things of this world). The Church had allowed itself to be entrapped by temporal sirens and had forgotten Christ's supreme law of charity and its own transcendental aims. In short, Luther accused the Church of neglecting its role of guide in the conquest of a heavenly kingdom.

The reasons for Luther's disapproval of the slender volume are more obscure, since the *Summa angelica* itself is obscure and would undoubtedly seldom be remembered now had it not assumed an emblematic quality on the fire at Wittenberg.[5] It is a handbook for confessors, for priests who administer the Sacrament of Penance, compiled at the end of the fifteenth century by a

Franciscan friar from Piedmont, Angelo Carletti (hence the epithet *angelica*). By burning the book, Luther sought to condemn a sacrament whose origins were medieval and which he viewed less as a means of purifying the faithful than of controlling them, since it placed the capacity to absolve the penitent's sins exclusively in the hands of the priesthood.

As I have already made clear, the motivations behind the Church of Rome's decision to enter into the legal sphere were complex, and we have considered them in detail above. What is important here, however, is Luther's drastically negative assessment of those motivations. Because of that assessment, the new religion of the early modern period withdrew completely from the legal sphere and concentrated instead on the immediate, internalized dialogue between the individual subject and the divinity. In addition to its rigid condemnation of any form of canon law, the Reformation also made a significant contribution to the more general decline in the prestige of the law and of lawyers, who were now seen as wielders of abstruse quackery who thought nothing of hoodwinking the unsuspecting citizen. The new religion also makes a substantial contribution to more general individualistic trends, and hence to the legal individualism that, as we shall see below, was beginning to spread through Europe, and would soon be the dominant outlook.

There is one final point to note. There is no doubt that Luther himself, who was after all a monk, was moved by a sincere belief in the need to purify an ecclesiastical establishment that had become deeply corrupted and that his ultimate goal was to liberate his new church from the suffocating embrace of temporal matters. However, once the Catholic Counter-Reformation was in full swing and the prospect of a papal *riconquista* had become real, the Protestant movement adopted positions that, although necessary for its survival, were in stark contradiction to the high-minded ideals of religious reform on which it was founded. The guidance and protection of many of the fragile reformed communities was entrusted to the national sovereigns who had joined the new religious movement. This led to the rather tawdry situation of monarchs heading national churches and the installation of the reformed denominations as official state religions. To this day northern Europe retains the relics of this antinomian politico-religious settlement.

Legal Humanism and its Two Souls: The Rational Soul and the Historicist Soul

Humanism represents a new way of looking at the relationship between the subject and the world: it is a revolution in anthropology. As such, it cannot but insert itself into every form of intellectual exercise, affecting the most profound choices made in such endeavours. It is unsurprising, therefore, that the study of law feels the effects of humanism.

The broad and fruitful cultural movement covered by the curt term *legal humanism* amounts to the synthesis and application of a variety of humanist premises. The reader should be aware that this was the most productive smithy for the forging of new ideas in all the early modern period. There are two considerations to bear in mind. Firstly, legal humanism was a moment of radical departure from the old certainties of the Middle Ages. Legal humanism is therefore not the final page of an old story, but rather the first page, or one of the first pages, in a new one. Secondly, legal humanism's influence is not restricted to the chronological era in which it manifests itself – the fifteenth and sixteenth centuries; rather, it has a decisive say in shaping the historical development of subsequent centuries.

In simple terms, the message of legal humanism can be reduced to a polemical attack on the perceived methodological deficiencies and consequent lack of rigour of medieval glossators and commentators and the brazen manner in which these figures were seen to have manipulated Roman law. Medieval jurists were happy to analyse Justinian's *Corpus iuris* minutely, but they never worried about the fact that the text represented the final result of a millennium of legal scholarship. They therefore did not think to compare the various stages through which the text had gone and which had been lost or suppressed by the sixth-century Byzantine chancellery in which the final version was written.

The emblem of these *homines novi* is the historicization of Roman law, a cultural battle which engages the most imaginative minds in Europe from the very beginning of the sixteenth century onwards: scholars such as Andrea Alciato from Lombardy, Guillaume Budé from France and Ulrich Zasius from what is now Germany provide models for new methods of legal scholarship that begin to form a widespread consensus.

Historicization meant seeing Roman law as an expression of and a mirror for its time. It was an ambitious project which demanded the incorporation of the study of law into other fields of learning: religion, philosophy, linguistics, literature, archaeology, politics. The humanistic scholar therefore needed a greater cultural openness and a broader education: a daunting requirement but a very fitting one for the men of the Renaissance, who were gripped by the ideal of *pansophy*. Pansophy is an obscure term which refers to the idea of universal or encyclopaedic knowledge: it described the humanists' vision of wisdom as the coherent sum of all knowledge, requiring the scholar in search of it to cultivate more than one restricted field of learning. Any intellectual discipline that attempted to proceed in solitude would stand accused of clipping its own wings and compromising its own integrity, because all knowledge was seen to be connected and pansophic. This was the reason behind the humanists' choice of philology as their defining cultural exercise; the practice transcended dry grammatical exegesis and became, as Budé would have it, a *doctrina orbicularis* – a plural and open form of study that allowed the humanist to divine the secrets of an entire historical civilization.

The humanists thus integrated other fields of learning as cultural props to their study of law. This contribution was fundamental to the structural reform which allowed Renaissance jurists to bring order to the chaos of Roman law. This order was certainly not the same order as that desired and enacted by Justinian – instead the previous readings of classical and post-classical law were re-immersed in their historical context. The new approach was to restore Roman law to its classical state and to follow faithfully back the changes provoked in Roman jurisprudence by the passing of a millennium.

This new methodology was a radical departure from that of the commentators and glossators and touched all aspects of legal study in the early modern period. In the humanists' desire to establish the authentic historical features of the Roman experience of the law, we should identify, and emphasize, two very different attitudes that tend to value different aspects of that historical experience. Both attitudes prove fertile contributors to the later developments of early modern legal thought. The first might be termed *rationalist* and the second *historicist*. Let us examine each in turn.

1. One of the 'betrayals' Justinian committed, as the humanists saw it, was his misguided use of fragments of classical legal wisdom; these were devalued and deprived of their authentic historical message by the emperor's 'manipulations'.

Classical Roman jurists came to be seen as sages similar to the logicians and geometers of ancient Greece in their capacity to construct rigorous and systematic theoretical models, the single elements of which were attributed the indisputable rigour of geometrical figures. The humanists therefore valued the perfect logic which the jurists used to construct their works. This logic could be perfect and incontestable because it is founded not in artifice but in human nature, what the rational humanists call *recta ratio* ('proper reason'). The classics were held to have read nature, to have understood it and then to have translated it into a laudable system of extraordinary logical cogency.

Roman law was seen by these scholars as *ratio scripta*: the written manifestation of pure rationality, which needed to be restored in order that it might serve as a model to the new Europe of the sixteenth century. The humanists did not see logic as dry and formal, but as a series of forms of reasoning that adhered to the real processes of thought. In their view, the rediscovery of these rational thoroughfares was vital because the systematic understanding of their work and its unifying principles depended on the re-establishment of logic. This intellectual environment favours the pansophic culture. Examples of this culture include the *De architectura* by the Roman writer Vitruvius and the *De re aedificatoria* by the Florentine humanist Leon Battista Alberti – this latter author was a well-known builder of churches and palaces but the work in question demonstrates that he was also true pansophist.[6] Budé, moreover, calls for his colleagues to style their works as *iure architectonico* ('on architectural law'). Meanwhile, philology, as we mentioned above, has become more than a simply grammatical and exegetical activity, and is rather a process which engages the rational faculties.

Given this pansophist outlook, the law's status as the ordering mechanism of reality becomes accentuated, thanks to the discipline's rigorous basis in logic. The logical bases of the law become even more apparent to the liberal scholars of the Renaissance after their study of the mathematical and geometrical architecture of the cosmos.

As will become clear below, this account of the rationalistic spirit of Renaissance humanism will be an indispensable aid when we turn to study the doctrine of natural law.

2. The pansophist current in early modern thought is also central to the historicizing spirit of legal humanism. I should reiterate that pansophism does not imply a confused mixing of diverse disciplines, but rather the use of a variety of disciplines to enhance and inform the detailed study of one field of learning by lending it methodological rigour and intellectual integrity.

The perception of Roman law as a historically generated phenomenon demanded that scholars contextualize Roman jurists' texts and techniques within a global historical account of Roman civilization as that culture had evolved over its various phases of development and in the various cultural disciplines practised by the ancient Romans. This information fed into the study of political and social history, religious history, linguistics, archaeology, epigraphy, and so on. The early modern jurist or legal historian studied these disciplines, not for love of eclecticism, but in order to develop a rigorous method of enquiry.

This leads, of course, to a flowering of historical and philological research on Roman law sources. So erudite are many of these jurists' readings that they are still useful to, and used by, contemporary students of Roman law.

However, this aspect of humanism is not so interesting to the modern legal historian, since we wish to follow humanism's effect on the subsequent developments in the field of law. We are therefore much more interested in establishing if there are any historical consequences of this apparently innocuous scholarly work on an enormous body of Graeco-Roman sources. The answer is strongly affirmative, because the restoration of Roman law to the Romans and the understanding of the discipline as a product of history are anything but innocuous.

In fact this operation forms the premise on which two significant developments are based: the differentiation of past and present, and the identification of the present as a specific epoch. Let us take the illustrative example of François Hotman (1524–90), a French historicist jurist immersed in the new currents of legal thought and a believer in reformed Christianity, as was often the case with such scholars. Hotman wrote polemical pamphlets addressed to the French king's government, full of concrete proposals for reform. These are all characterized by a desire to further the kingdom's autonomy: Hotman argues that the study of Roman public law is futile since the demands it sets out to satisfy are so radically different from those of sixteenth-century France. Hotman bases his ideas for reform on ancient Frankish institutions – a seemingly naive move, but one that demonstrates his desire to find indigenous French historical sources that will grant the nation legal autonomy. He explicitly calls on the king to set up a unified national legal system based on French regional customs.

The first principle that Roman law is a historical artefact premises a second, more general, methodological assumption: legal systems are all historical artefacts and, as such, deserve to be evaluated along with the historical moment they reflect. French law is thus equally a mirror of the society and customs of its own historical period. Historicist Renaissance scholars do not shy away from making such evaluations; they establish themselves at the heart of a nation-state that is constructing, little by little, its sovereign independence. The humanists throw themselves into the construction of French national law, devising projects that are immediately applicable to its needs and insisting at times on using the regional *coutumes*, since they see these unrefined sources as the best expression of the spiritual unity of the French people, and at other times on the need to strengthen the legislative powers of the French king.

Ironically, these antiquarians turn out to be the thinkers who establish anew an intense feeling for the identity and respectability of the present moment; in so doing they also turn out to be some of the strongest allies of the king. The nascent French legal nationalism draws significantly on this humanistic current of thought in establishing itself.

Natural Law

If legal humanism was the prime mover in the study of law across all of Europe and especially in France during the fifteenth and sixteenth centuries, the seventeenth and eighteenth centuries see the whole of the continent in the grip of a new movement: the doctrine of *natural law*. I shall now seek to help the reader understand this important and extremely consequential historical trend.

At its simplest, the movement entailed premising the legal culture of the age on the notion of natural law. The doctrine of natural law's defining feature is an internal dialectic between the assorted positive historical forms of law produced by various political bodies and a higher form of law – natural law itself – which is not produced by those political bodies but exists above them in a superior, universal reality, which, for now, I shall call *nature*.

The recourse to natural law is certainly not exclusive to the early modern period: Graeco-Roman culture, medieval culture and even our own contemporary culture, have all participated in a similar dialectic to the one just outlined. We can, in fact, identify a common and unifying set of qualities in these recourses to natural law over the passage of time: natural law tends to be liberating and emancipating. It often sets up a series of principles and rules in opposition to the contingent and arbitrary choices made by the wielders of power, meaning that the meanness of the powerful is counteracted by something more durable and solid.

Having said that, one should immediately add that we must not be satisfied with such a generic account of the question, since it does not give the full picture of the early modern doctrine of natural law; greater historicization is still needed. Let us begin with two clarifications: in seventeenth-century Europe, an entire legal order grows up based on the dialectic between natural law and positive law; furthermore, a completely new meaning is attributed to the generic phrase 'natural law', which has been variously understood at different points in history.

Since comparison is the keenest weapon available to the historian, let me call upon the medieval view of natural law. Thomas Aquinas, who, as ever, gives a faithful and lucid account of medieval thought on the matter, bases his analysis not on the figure of the individual but on that of the cosmos. The universe is seen as magnificently ordered for mankind by the rules and principles laid down generously by God and inscribed in indelible characters upon the things of creation. Here natural reason is divine gift with a rigorously *objective* connotation.

But between the end of the thirteenth century, when Thomas was writing, and the beginning of the seventeenth, when the doctrine of natural law was at its height, run the liberating currents of the sixteenth century, already dealt with in depth above. Humanism, proto-capitalism and the scientific revolution are seminal events which all combine to overturn the previous view of the natural and cultural worlds as oppressive. These revolutions are the intellectual prerequisites to the seventeenth-century doctrine of natural law and they lend it its novel and epochal qualities. Natural law is forged in the historical crucible of change that is the Renaissance – beginning in the fourteenth century and coming to maturity in the sixteenth. It participates fully in the revolutionary nature of the new anthropology of the era. I shall now attempt to sketch a basic outline of the natural law movement. This must obviously be seen as flowing from and cohering with the preceding current of liberation of the individual subject, with respect to which the doctrine of natural law functions as a further step forward that lends the prior trend indisputable validity.

In the seventeenth century, the mathematical and natural sciences now provide a secure methodological fortress for jurists, who now claim to make *discoveries* like the natural scientists. In the same way as someone who discovers a universal physical law that governs the way things naturally behave, jurists seek to establish universal rules of human conduct as defined by the nature of mankind – something which should be visible to anyone who has the correct outlook.[7] One must therefore find a means to *discover* true human nature, which is not something easily done because the passage of history has deformed and misdirected that nature.

This problem gives rise to the issue which most troubles the acolytes of natural law: the question of how to liberate oneself from the artificial impositions of history and arrive at a perfect understanding of what the primitive natural state of mankind was. The first human is therefore seen as an uncontaminated model that may help scholars recover the nature of his race. This preoccupation with primogeniture may seem like a naive and harmless myth, but its importance becomes apparent if one notes the form this *human nature* takes: a primal state in which each individual subject was absolutely free to govern his own actions.

The most basic natural law is the instinct to safeguard the individual self, which leads to the only existential duty: self-preservation. Humanism's anthropocentrism here transmutes into the most rigid form of individualism: in the natural state, the individual is not bound by any social or collective ties – these are later impositions when groups of many individuals, each motivated by self-interest, choose freely to create a civil society. The world of *history*, in this view of humanity, exists in Manichaean contrast to that of *nature*, just as purity is in opposition to contamination.

In order to illustrate this point of view, let us take the example of the Dutch jurist Huig van Groot (1583–1645). Van Groot (or Hugo Grotius, to give the Latinized name by which the jurist is often known) is the first writer to give a global account of this new legal anthropology, principally in his long and theoretically sophisticated work the *De iure belli ac pacis* ('On the Law of War and Peace', 1625), one of the seminal works of early modern legal thought.[8]

In the work's extensive preamble, van Groot gives an excellent account of the new legal outlook: 'first of all, I have been concerned to base my proofs regarding natural law on notions so evident that no one could deny them without committing violence upon themselves. In fact the principles of such a form of law [...] are manifest and evident in and of themselves' (*Prolegomena*, XXXIX). Here we can see for the first time a concern with *evidence*, a term which will be a consistent refrain of the seventeenth century. The emphasis on evidence makes clear that we are dealing with a cultural context that has undergone a complete sea change since the Middle Ages: what before was seen as a gift from God, human reason, is here isolated from its natural and cultural surroundings, and evidence serves as the fulcrum that allows the subject to pivot towards the heights. Evidence does not require divine intervention, but functions as its own justification. The new doctrine of natural law is rigorously secular.

Van Groot continues: 'I can declare openly that, just as mathematicians contemplate shapes by making abstractions from distant bodies, so I, in my treatment of the law, have abstracted my thoughts from any instances of fact' (*Prolegomena*, LVIII). Again we see the discoveries of legal scholarship associated with a new and more satisfying mode of research. Van Groot, who was an attentive student and admirer of Galileo, wholeheartedly adopts the mathematical sciences as his epistemological model; he bows down before them, acknowledges their primacy and takes every opportunity to cite their example. Jurists like Van Groot thought they could foresee a moment in which they would be able to reduce the complexities of their subject to a harmonious series of rigorous geometrical figures. Before long, legal scholars were talking of a *mos geometricus* ('geometrical rule'), a type of law which is precisely opposite to the chaotic sedimentation of laws that characterized the medieval legal order. The law was to be simplified to its most essential outlines, as in the theorems and postulates of a geometer.

But in the second phase of van Groot's argument there is another significant innovation that needs to be emphasized: van Groot denies the factuality of the law that had been the central tenet of medieval legal thought, even in its scholastic phase. Instead, following the 'true' science of mathematics, van Groot opts for an abstract form of knowledge. Nature is merely a laboratory in which the jurist may experiment and even fantasize. The seventeenth-century doctrine of natural law is thus a process of reasoning through models – a grand and elaborate discourse predicated on an abstract vision of humanity that has never had concrete existence because it is not made of flesh and blood but is rather suspended in a sort of earthly paradise outside of space and time.[9]

But why construct a mythical world? Earlier I asserted that the myths in question were far from innocuous and that assertion still holds; now we must examine why such myths arose and what they mean for our argument.

The seventeenth-century doctrine of natural law is complex and in some ways ambiguous. It is without doubt a great intellectual venture, but its originators are certainly not independent of their historical context; rather, they are immersed completely in that context and attempting to bring some order to it. Their emphasis on the pure state of mankind serves to distinguish a sphere in which the liberties of the human subject cannot be constrained by the arbitrary interventions of power. As we shall see, both modern constitutional thought and the very first barriers erected to protect the individual – the seventeenth- and eighteenth-century bills of rights – stem directly from the intellectual venture of seventeenth-century natural law.

The thinkers who set up this doctrine of natural law are also men of their time because of their individualistic worldview, which can be seen in their incessant attempts to abstract the individual from any social organization and to assert the value of an essentially selfish attitude. Van Groot's idea of the natural state is underpinned by one fundamental rule: the respect of every person's *proprium* (literally 'own things'). Van Groot believes in an independent individual sphere in which rights of liberty and property are found alongside one another. This conclusion is even more pronounced in the account of a new type of society laid out in the English philosopher and political thinker John Locke's (1632–1704) *Two Treatises of Government*. Written between 1680 and 1690 – a portentous decade in English history – the *Two Treatises* describe the state of nature as one of 'perfect freedom', in which the role of guarantor of individual liberty is attributed to *property*.[10] I italicize the word *property* here to make clear that it carries a specific nuance that is identical to van Groot's *proprium*: a complex of individual impulses and abilities. *Property* has an enormous capability to expand and project itself onto external things, which it then brands with the mark of exclusivity, tying them to an individual subject as if 'me' and 'mine' were interchangeable.

And this is the ambiguous aspect of the early modern interpretation of natural law: from one point of view it is an intellectual venture undertaken according to the objective rules of scientific research, and yet it is deployed very subtly by thinkers whose needs and interests align with those of proto-capitalism, already a thriving reality in colonial powers such as England and the Netherlands, in order to set up a new socio-legal reality in which those needs are satisfied and those interests safeguarded.

The new individual, around whom society pivots, is the direct heir of the emancipated subject freed from the trappings of the Middle Ages whom we saw emerging in the theological and philosophical developments of the fourteenth century. The individual is a free and liberated subject thanks to his status as *dominus*, or proprietor, both of goods but above all of himself. The journey which would find expression and fulfilment in the second of Locke's *Treatises* was thus embarked on in the by now remote fourteenth century. Yet the central thesis remains the same: the ownership of goods is seen as *natural* since it emanates from the ownership of oneself and the instinct for self-preservation, which is instilled by the deity to safeguard the self and is therefore indisputably *natural*.

The subject that receives such attention and care from early modern thinkers is a property owner; private property is thus at the heart of the new order. The new society can rightly be termed *bourgeois*: the ascendant class is the bourgeoisie, which is motivated by profit and the accumulation of wealth. The bourgeoisie assumes its central role by fusing the realms of ethics and economics, elevating the act of possession to a higher level previously unheard of. Personal property, in fact, becomes identified with an individual's selfhood and inhabits the most intimate recesses of the subject, becoming the most significant contributor to a person's individuality. During the course of the eighteenth century political thinkers, economists and jurists will not hesitate to term personal property 'sacred' – a designation echoed by Article 17 of the first French *Declaration of the Rights of Man* in 1789.

There are two issues in need of further elaboration: the preference for abstraction and the role of political power.

We have seen above how the early modern doctrine of natural law deals in abstract subjects, that is to say with improbable beings who inhabit a non-existent space and time, that of *nature*: not flesh-and-blood people so much as models of people. But we must ask how this preference for abstraction can be reconciled with the privileging of riches and the central role played by wealth, which necessarily lends a bitter air to the very concrete divisions between haves and have-nots.

Because of the privileging of abstraction, the doctrine of natural law marks the beginning of a strategy that forms part of bourgeois law throughout modernity. Subjects, relationships, freedoms, equalities and rules are all described in the abstract, yet this abstraction constitutes a very effective figleaf placed over the concrete aspects of a society riven by injustice. Abstract affirmations succeed in representing the action they describe in formal terms but prove completely unsatisfactory in the social sphere. One glaring example is the application of freedom to contract law and the formal equality of both parties who enter into a contractual agreement. In theory the lowliest beggar ought to benefit by being held legally equal to a well-fed burgher, but in reality the equality is only nominal and does little to stop the richer party crushing the poorer. Abstraction thus principally serves those who own property, since the rights it grants remain virtual, and the economic and social disadvantages of the weaker party are left unaffected.

There was one very significant change which resulted, however. The privilege and exclusion which attached to social class were eradicated and overbearing corporative structures were torn down (or at least attempts were made to accomplish this). The individual now stood alone in front of political power, or rather a populace of individuals did so.

Let us swiftly deal with the second point for clarification: political power is seen in a different way in the era of natural law. According to the doctrine of natural law, power is not only *natural* but *historical*: people consent to form a political society because they have self-interested motivations for doing so. The attitude towards political power is that it must safeguard one's *property*. The political structure must therefore maintain public order and safety by means of an effective police force. Individual property owners generate political power and that power is bound to protect them. The bourgeois era is thus characterized by a reciprocal amity between the bourgeoisie and the political class.

The Legal Enlightenment: Legalism and Legal Idolatry, the Age of Legal Absolutism

The final reflections of the previous paragraph regarding the doctrine of natural law's fluctuation between ambition and ambiguity make an excellent preamble to our analysis of the European legal Enlightenment. The legal history of the eighteenth century can be seen as the continuation, the fulfilment and yet the betrayal of the endeavours of the natural lawyers. Like the doctrine that preceded it, the legal Enlightenment was a current of thought and actions on a continental scale. Indeed, it can be argued to be a continuation of the doctrine of natural law, with certain characteristics accentuated and greater specificity of thrust; however, these alterations meant that the lawyers of the Enlightenment reached solutions that were diametrically opposed to the premises of natural law.

Let me make one initial clarification, necessary for the reader to understand the legal Enlightenment correctly: whilst the natural lawyers of the seventeenth century showed a great determination to translate their ideas into a reformist project affecting the social, political and religious world, the Enlightenment lawyers of the next century were even more determined in that regard. In the specific area of this study, the law, they sought to enact a nakedly political view of the law, attacking head-on the nexus of relationships between subject, nature and political power that was undoubtedly present in the seventeenth-century legal project but far from fully implemented in law.

Hence one of the first qualities of the legal Enlightenment: the movement represented a broad community of intellectuals who had gained emancipation thanks to the cultural revolutions of humanism and natural law and were now convinced of their ability to read the natural and social world correctly and of the capacity of their insights to align historical reality with the recently identified natural laws. This community includes monarchs and politicians: it is a community that plans, attempts and sometimes succeeds in enacting the structural reform of society and of the apparatus of the state. A new matrix of ideas and procedures for political and legal activity was instituted across Europe. Examples of the sea change in governance of the eighteenth century include King Frederick II of Prussia (1740–80); Empress Maria Theresa of Austria (1740–80) and her minister von Kaunitz (and subsequently Maria Theresa's sons Joseph and Leopold); and King Joseph I of Portugal (1750–77) and his minister the Marquis of Pombal.

A new idea of the monarch came about: sovereigns now delighted in entering into cultural disputes and in presenting themselves as generators of legal thought. For example, in 1750, Frederick the Great presented a *Dissertation on the Reasons for the Establishment or Abrogation of Laws* to the Berlin Academy. Legislators were also different: they felt obliged to make their legislative projects represent a manifesto on behalf of the new philosophy. The Marquis of Pombal, for example, drafted a document called the *Lei da Boa Razão* (the 'Law of Good Reason'), which demonstrated the penetration of the principles of natural law, given that they had reached chancelleries in the farthest western extremity of the continent.[11]

The legal Enlightenment represents a truly political view of the law: it tackles the problem of the relationship between natural law and political power and resolves it via an innovative reorganization of the sources of law. The law itself becomes the privileged object of intellectual reflection and political action, something which rocks the discipline to its core, bringing to it a new absolutist dimension.

The need to guarantee the freedom of the individual and his property had already become clear from the advances of natural law. Locke himself, who had championed the most liberal interpretation of such guarantees, had put forward a very legalistic view in which the freedom of subjects was tightly linked to the laws passed by the prince. However, it was not until the legal Enlightenment that the political visions of the seventeenth century became first proposed and then implemented reforms. The protagonists of the Enlightenment are no longer philosophers and political thinkers but jurists and economists (the latter a new entrant on the European intellectual stage).[12] The process has two parts: firstly there is a destructive critique of the of European law's customary sources; secondly there is a series of constructive proposals.

By the mid seventeenth century the medieval system of law was becoming a burden to the whole of Europe. It had been eroded somewhat by the ever greater incursions of sovereign, authoritarian laws passed by strong nation-states (especially, as we shall see, in the kingdom of France). However, a great morass of customs, doctrinal opinions and judicial rulings still remained in force and, with the passing of the centuries, this gave rise to a chaotic and haphazard legal environment in which uncertainty was the rule. In such an environment, primacy could be granted and was granted to the judgments of jurists – the possessors of technical knowledge. The confusion, obscurity and indecipherability of the eighteenth-century legal system clashed with the need for geometrical clarity, certainty and simplicity which was asserted vigorously by post-humanistic legal thinkers.

The legal system was thus viewed as without redemption by both the monarchs of Europe and Enlightenment intellectuals. The sovereigns despised it because it offered too little control to those in power; the intellectuals disparaged it for more complex reasons, of which more below. The thinkers of the Enlightenment had little regard either for the customs or for jurists. We should not forget that eighteenth-century intellectuals tended to be aristocrats, often belonging to Masonic lodges or other exclusive groupings. The idea of a system of law which grows from the ground up, from the complexity of human activity, was thus foreign to their way of thinking. For example, the works of the Frenchman Charles de Montesquieu (1689–1755) and those of the Italian Cesare Beccaria (1738–94) represent two very diverse standard bearers for the European Enlightenment. In both cases there is a similar attitude to the common people, for whom both authors show disdain, dismissing their opinions and anything that they may contribute. Jurists receive scarcely better treatment: both authors condemn them as too technical and not philosophical enough, as too detached from the culture of their time and too wrapped up in the common law, which they see as compromised by its medieval origins and its association with the Church of Rome, an institution they abominate, and with the canon law. The process of legal reform thus had to begin with a drastic repudiation of the past, in order to clear out the old sources of law and eliminate them as if they were a source of shame.

The void was very well filled by supporting to the hilt the claims to political autonomy of the new nation-states, where monarchs were now taking on the role of legislators and beginning to create national systems of law directly controlled by them.[13] There was therefore a great degree of understanding in eighteenth-century Europe between the holders of political power and the *homines novi* of political and legal thought. There was also a massive increase in the role of the prince in the area of the production of law, a role that soon became idealized.

As the men of the Enlightenment saw it, customs were, by their nature, hidebound by the facts with which they dealt; judges could not avoid being immersed in the particular conflicts they were assigned to resolve; schoolmen in departments of law could not lift themselves above the technicalities with which their knowledge dealt; only the prince, in his ideal role as the model of mankind and champion of all virtues, could remain untouched by human emotions and accomplish the objective contemplation of the common good necessary for maintaining public contentment. The prince was therefore the only figure to whom they could accord the responsibilities of governing, of policing society and of creating new laws.

Who better than the prince, with his lofty viewpoint, to separate natural principles from artificial creations, to interpret the nature of things and translate it into laws? In the secularized culture of the eighteenth century, the former monopoly of the Church of Rome as absolute guardian of the natural law is now placed in the hands of the prince. This leads to the very significant consequence that the true law, the law that supposedly derives from the natural geometry of society, is now seen as being identical to the will of the prince and expressed exclusively by his laws.

In pre-revolutionary France, the physiocrats, the first to sketch out an economic analysis of society, talk unconcernedly of a 'legal despotism', secure in the knowledge that the prince interprets what is *evident* and translates it into general laws.[14] In Italy of the mid eighteenth century, two ringleaders of the Italian Enlightenment, Ludovico Antonio Muratori (1672–1750) and Cesare Beccaria, assign the prince a decisively absolutist position in two fiercely polemical books. Muratori's *Dei difetti della giurisprudenza* ('On the Faults of Jurisprudence', 1742) is rigidly explicit: 'the most laudable result would be if the Princes were to take a scythe to the root of this problem, cutting short all controversy and bringing inviolable order to everything that should in future come before the courts of justice with new laws and statutes'.[15] The reason is simple: 'if Princes write the laws, the only thing they will have in mind is the public good' – a quotation that highlights the astounding process of idealization of the role of the prince that had occurred in European Enlightenment legal thought.[16] Beccaria also comes to the same striking conclusion, using a very eloquent formulation: 'the monarch, or rather the representative of the current will of all'.[17]

The most serious problem is that, when these writers talk about princes, they do not mean a universal prince (the position of Holy Roman Emperor, which continued to exist for a few more years, was by this point little more than a phantom), they mean a fragmented plurality of sovereigns each of whom controlled a stretch of territory defined by divisive borders. What a narrow outlet for the torrent of universalism that had so boldly underwritten the doctrine of natural law! When earlier I defined the Enlightenment reflexes of the doctrine of natural law as a betrayal of its principles, I was referring to examples such as this. In accordance with the basic ideas of a secular civilization, the European Enlightenment had made the holder of supreme political power into the interpreter of the natural law, but, in so doing, the universal message became particular and the position was given to the heads of the individual nation-states. This shift from universalism to particularism lies at the basis of modern political and legal systems and represents the pre-eminent contributing factor to the powers of the individual nation-state, since it lends them an ethical legitimacy derived from the inherent ethics of natural law.

Obviously, since natural law and the legal Enlightenment were as much political strategies as they were philosophies of government, the politicians of the time devised much clever propaganda to obscure the profound contradictions that cannot help but strike the historian. A whole mythological edifice, founded on undemonstrated beliefs, was thus erected to defend the new system.

One pillar of this edifice was the above-mentioned idealization of the prince so as to make him a trustworthy model. We must now examine another pillar of this edifice which supported and complemented the first, namely the substance of the law itself, that is to say the document which makes clear the prince's will and asserts his power – in other words, the means by which the prince's authority is projected over the community of his subjects. A very subtle campaign of propaganda is waged in fervent praise of the law. The law, which emanates from a model, from a symbolic figure, has none of the hatefulness and subjectivity of a command issued by a person of flesh and blood. Instead the law is seen as the expression of 'human reason' (as Montesquieu would term it) or 'general will' (to follow Rousseau's formulation). Obeying the law does not impact on citizens' liberty, since freedom consists exclusively in security from violence by others: a security that only the law, with its general, abstract and rigid commands, can offer. The abstract and general nature of these commands is held to make serving the law the opposite of serving a man. 'Public happiness and the veritable good demand that government by men cease and government by laws begin and that the sacred power of making laws be kept jealously by the crown and nowhere else'.[18] This is a message which reaches its fullest expression in the work of Rousseau (1712–78): people are free where they submit to the law, not where they obey another person; obedience to the law is obedience to the public will; the law makes a people free by expressing the people's will.[19] This set of beliefs is well fashioned, but it is predicated upon an assertion of coincidence between the written law, which represents the will of the office invested with supreme power, and the will of the national public.

Behind the rhetorical trappings lies an authoritarian impulse, and the eighteenth century sees a swift and dangerous slide towards authoritarianism. This is a message that emerges surprisingly clearly from the last great Enlightenment thinker, Immanuel Kant (1724–1804). In his 1784 essay 'What is Enlightenment?', Kant emphasizes the coercive nature of laws and the absolute necessity of obedience. He returns to this theme in 1797, in the *Metaphysics of Morals*, which contains a section dedicated to the theory of law and the impermissibility of any opposition by the people to the state as supreme legislator.[20] Behind the figleaf of 'natural reason' and the 'general will' lies the deposition of the whole mechanism of the production of law in the hands of the politically powerful. The old chaos of legal pluralism is replaced by an extremely rigid legal monism: the law is now bound to the apparatus of state power and tends to become conflated with legislation. And so begins a long period of legalism, and even of legal idolatry: the law itself, as the expression of a sovereign will, becomes an object of worship whose content is unimportant. This profoundly dubious attitude to the law continued to be very influential until the very recent past, indeed only in the last few decades have we begun to rid ourselves of it.

From the eighteenth-century Enlightenment onwards, the legislator held sway over the European continent; he was the only figure who could make any given social rule into law. As a concluding point, I should also point out that the culture of economic liberalism does not hesitate to don the armoured suit of legal absolutism in its attempts to gain control of the law, which is so integral to the social order.

The Age of Proprietary Individualism: Economic Liberalism and Legal Absolutism

This section will be somewhat more concise than the previous one, although not because its content is less important – quite the contrary. However, the important issues of this section have been treated above, albeit in a somewhat sporadic manner. Let me recapitulate for the reader's convenience.

The beginnings of modernity coincide with those of a proto-capitalist economic and anthropological system. The beginning of modernity is marked by the all-conquering rise of the bourgeoisie, which finally gains political ascendancy with the French Revolution at the end of the eighteenth century. As we have already seen, economics acquired a new importance, because a completely new meaning was attributed to property, which was now seen as the external projection of the subject. Property was no longer an appendage with connotations of riches and bodily comfort; it was something greater – in the modern society, he who *had* more *was* more. Property was also generalized beyond things: it became and sought to become Locke's *property*, the central expression of the internal reality of the subject, interchangeable with a person's liberty.

Property owners were thus the centre of attention, because they were at the heart of the public sphere. Owners of property were seen as *virtuous* subjects, quiet and disciplined and therefore a suitable foundation on which to base the continuance of the new political order, which had begun to establish itself, first in England and then, after the revolution, in France and the rest of continental Europe. The new movements wasted no time in setting up the most formidable strategies to defend themselves: property was construed as part of the natural state of the earthly paradise: the first man was seen as a property owner and, moreover, as engaged in discharging the only true, sacred duty, that of self-preservation. Individuals' property thus becomes sacralized.

The late eighteenth century is a time of amorous declarations between the bourgeoisie and the new culture, who found themselves united in an alliance against aristocratic and clerical privilege and the deluded vestiges of an anachronistic feudal system which the decrepit governments of the age still sought to uphold. The legal innovations of abstraction and generality seemed tailor-made for a bourgeoisie, and the celebrated new legal principle of equality was also very favourable to the middle class since it ignored differences in wealth. The idea of a strong, unified and robust state with a firm grip on the production of law, by now entrenched across Europe, was a further advantage.

The middle classes, in their aim to conquer political power, could see, with lucid prescience, that only a powerful public sphere would allow them to secure total and guaranteed sway over economic matters. The state suits the bourgeoisie. Economic liberalism and legal absolutism are only apparent opposites: on the contrary, the symbiosis of the two soon becomes very successful, as can be seen the extraordinary increase in the power of the legislature and of acts of legislation, especially on the continent, and the codification of all law which soon follows the rise of the middle classes (see below, pp. 84–87).

The Age of Constitutions: Modern Constitutional Thought between Myth and History

Modern constitutional thought is born out of and built upon the doctrine of natural law's need for reflection and consolidation. It seeks to place curbs on political power and to protect the individual via the fostering of unrepressed expressions of subjectivity. This is most certainly the beginning of a historical current that will lead to our contemporary situation, where we rejoice in a refined legal culture in which the fundamental rights which we now see as our inalienable political and ethical inheritance are well established: the right to freedom of religion, to freedom of speech, to freedom of movement, to freedom of assembly, and so on. However, I should repeat that, alongside the rights that maintain the intellectual and moral integrity of the subject emerged a situation that also favoured the interests of the property owner because of the belief in individualism and the fusion between being and having. The doctrine of natural law was the cornerstone on which the bourgeois order was founded.

The first manifestations of constitutional thought are to be found in certain written documents, which are usually known as *bills of rights*. This blanket term does express appropriately the fundamental unifying ideal that links these documents, but we should not let their differing historical characteristics go unremarked.

The first bill of rights is usually taken to be the 1689 English Bill of Rights, signed a year after the Glorious Revolution and the departure of the Stuart monarchy. The Bill's importance should not be over-emphasized: it represents merely a set of precise limits for royal power and an invocation for the monarch to respect certain rights of the subject that had been violated in the recent past. The document is clearly born out of the end of a political struggle and represents nothing more than a hard-won historical victory.

If the Bill of Rights is in any way a 'constitutional' text, it is so only in the most embryonic manner. The constitutional aspect will become far more overt in the bills of rights and constitutions of the first North American states as they seceded from Britain (1776, 1777, 1780, 1784) and in the *déclarations*, *constitutions* and *actes constitutionnels* of the six years of revolution in France.

We must not ignore the differing historical contexts of the two sets of texts: the American declarations are refusals to obey orders from the British monarch that are deemed arbitrary and despotic, whilst the French situation represents a comprehensive rejection of a whole social, political and legal structure grounded in social class and a post-feudal legal system.

Nonetheless, the American and French documents of the late eighteenth century emerge from a similar historical environment and give complete expression to the message of natural law. They aspire to be a reading of the nature of things and of the principles and the supreme laws written into those things; they seek to declare the fundamental legal rights of subjects in their natural state that those with political power must respect once the community of subjects has decided of its own free will that it is opportune to form a political relationship.

The mythology of natural law thus forms the connective tissue of these foundational ethical and political texts. The installation of individualism and of the bourgeois hegemony described above is thereby confirmed, and these changes are boiled down to a catalogue of fundamental individual rights. It would be unfair to deride the genuine advances made in the modern period; one should certainly give strong emphasis to the formidable steps forward that were taken – the breaking of the shackles of class, and the recognition of subjects' rights to express and govern their own personalities. That said, one cannot help noting that the declarations do not go far enough, as I have argued above.

The original sin of natural law hangs over the bills and declarations here, which all seek to base themselves on a point of arrival/departure that claims to be a state of nature but is in reality an intellectual construct and, as such, supremely artificial. It is a state of nature populated not by persons of flesh and blood but by models of men, equal and perfect like casts from the same mould. They therefore create an idealized scenario, which rejects any need to verify itself in the concrete but often disappointing reality of history. History is dismissed as something that creates men but also deforms them, and yet it is surely ingenuous and misleading to reject it as the constitutionalists do.

As one might expect, these premises led to a catalogue of rights that is theoretically persuasive in its sonorous pronouncements of liberty, equality, rights and even happiness (a naive term that recurs frequently in eighteenth-century declarations). However, the list contains little to help the destitute members of the working classes, the misery of whose daily lives was left untouched by this swathe of reforms, which must have seemed irrelevant, decorative and even derisory to those in the grip of hunger. The state of nature hands down a static image of humanity, as befits the rarefied air of meta-history, but life as it is lived day by day is at the mercy of all the forces of history.

Virginia was the first of the ex-British colonies to adopt a Bill – the Declaration of Rights (1776). After the solemn opening affirmation 'That all men are by nature equally free and independent' the inalienable rights given are: 'the enjoyment of life and liberty, with the means of acquiring and possessing property, and pursuing and obtaining happiness and safety'. It is a mannered piece, like one of those precious bucolic landscape paintings in which shepherd boys wear white wigs like rich Parisians. The elite frame of reference is made clear by the assertion of a link between liberty and property, a link that will return in the French *Déclaration* of 1789, and will remain untouched in all the various documents that express the permutations of the revolution through to 1795. Hence the individual rights held to be fundamental are those of liberty, security and property. It must have been a very strange feeling for proletarians to discover that they were to be treated as potential property owners despite having effectively nothing to call their own!

And so declarations of rights float on top of an unpalatable, and therefore unmentioned, social reality, all the more so because alongside their thunderous declamations there is barely ever a list of responsibilities that might have helped to ground the rights upheld in the workings of society. It is only in the context of the corresponding duty that any right can find its relative social weight. The *Constitution* of the French Republic, of 22 August 1795, does follow the section on rights with one on responsibilities. But so scant is it that it is frankly a disappointment to read it. Such generic mention as there is of moral duty is made in the most hectoring tone reminiscent of the worst kind of religious paternalism.

So much for the bills of rights: they may be the first historical manifestations of modern constitutional thought, but they are constitutionalism without a constitution, as it were. The phrase may seem a surprising one at first sight, and in need of some explanation. How can there be modern constitutional thought without a constitution? Meanwhile, political historians often refer without a second thought to 'the constitution of the ancients', the 'medieval constitution', the 'ancient Constitution' of England pre-1688 or the 'ancienne Constitution de la France' before the revolution.

But let us be clear: *constitution* is a polyvalent word and can be understood generically to mean that inheritance of conventions and customs which inevitably accretes and solidifies over a long period of time in any political community. However, if we interpret the term more specifically and less generically, it means an organic collection of principles and underlying and overarching rules which a group of political officeholders have derived from the values of a people at a given moment in history. This collection of principles and rules is identified as the deepest and the most authentic legal signature of that people and is thus transformed into a complex and detailed document. Constitutions of this type come about relatively late, during the twentieth century, when the distinctive features of legal modernity give way to forces and values that indicate an arrival in a new legal landscape. But that is a topic for later.

In conclusion, bills of rights are testaments to a legal climate that was tending towards constitutionalism. However they represent, in my view, only proto-constitutions – bricks for an edifice as yet unbuilt.

The Kingdom of France in the Seventeenth and Eighteenth Centuries: The Making of French Law

In the pages above I have been swift to emphasize the attention paid by French monarchs to producing and controlling the law. We have seen how there was a progressive growth in royal lawmaking and how sovereignty becomes progressively closely identified with the power to legislate. But we have also seen how, at the same time, the French monarchs showed a remarkable prudence in their legislative activities, always holding fast to an awareness, which lasted throughout the Ancien Régime, that the king should never seek to overturn the accumulation of customs that had gradually formed itself into a sort of unwritten constitution – the fundamental unwritten laws of the realm. Even at the end of the sixteenth century, we saw that Jean Bodin, the great theoretician of absolute monarchy, still distinguishes between two different legal orders that govern different areas of France: *loy* – the discretionary power of the king – and *droit* – law that stems from customs. And in the seventeenth century Louis XIV, who is customarily cited as the epitome of absolutism, thoughtfully rejected the brave proposal of his minister Colbert to 'unify the legislature' – holding that the times were not ripe to carry out such a radical reform of the customs.

At the outbreak of the revolution, the complex legal landscape of the nation remains, therefore, substantially unchanged, although there have been some attempts by the kings of the sixteenth and seventeenth centuries to enlarge the portion of the legal sphere under the influence of royal legislative power. The most conspicuous of these attempts is certainly that carried out by Louis XIV and Colbert, via four penetrating declarations: the *Ordonnance civile* of 1667, the *Ordonnance criminelle* of 1670, the *Ordonnance du commerce* of 1673 and the *Ordonnance de la marine* of 1681. The four *ordonnances* constitute a legislative programme without parallel in other monarchies, as is evident if one compares them to the various *recopilaciones* promulgated by the Spanish kings between the sixteenth and the nineteenth centuries, for example.[21]

The first two *ordonnances* are usually known by the name *Code Louis* and the third by the name *Code Savary*. The term *code* here may provoke some misunderstandings and so let me put these documents in their proper historical and legal context straight away. *Code*, like constitution, has a number of meanings and can therefore be used in a number of different ways. The Latin word *codex* originally denoted the material, be it wood or paper, on which a text was written. In legal terminology, the word came to mean a collection which organized a diverse group of laws: hence the ancient codes, such as those of Theodosius and Justinian and their many successors in France of the sixteenth to nineteenth centuries.

However, when legal historians talk of the 'age of legal codes', and specifically of the 'Code', they are referring to a moment in the history of legislation characterized by a very particular type of legal source that should not be confused with these earlier experiments which we dub 'codes' for formal reasons because they result from a prince's or a jurist's attempts at a general organization of legal material. The meaning of *code* in the technical language of legal history assumes two preconditions: firstly the great cultural revolutions of the Enlightenment and of the doctrine of natural law; secondly an omnipotent and daring legislator with no compunction towards legal pluralism in the territory under his control, who seeks to bring the entire legal order under the aegis of his legislation. This type of code partakes of the legalistic and legally idolatrous myths of the Enlightenment in order to take on the role of an archetype capable of imbuing certain historically produced documents with a totalizing value. To group the Napoleonic Code (of which more very soon) with the *Code Louis* and the Justinian Code because of a superficial verbal resemblance is a form of nominalism that places the cart of terminology before the horse of historical meaning.

None of this changes the fact that Louis XIV's *ordonnances* are a very important quartet of legal-historical sources and deserving of our attention. The first three will be covered here. The fourth, published in 1681, is also interesting but deals mainly with maritime law, although this fact in itself demonstrates that contemporary France was already major colonial power with trading partners from across the world.

The first two *ordonnances* sought to unify and reform civil and criminal trials. Their reorganization of the judiciary had one precise legal and political aim: to redefine the role of the French parliaments. These had originally begun as judicial bodies, but were now beginning to erode the king's power by claiming the role of representatives of the nation and interpreters of its 'fundamental laws'. The *ordonnances* reiterated the subaltern status of all judges, including the parliaments, to the law, which was itself identified as coinciding with the will of the king alone. They moreover contributed to the reduction of the law to a system of legislation. These aims lend the two *ordonnances* their historical and legal significance. Although they limit themselves to ordering pre-existing legal material and do not claim to be the exclusive manifestations of the law, they were nonetheless a certain and decisive step forwards on the historical journey towards the Code because they used royal legislation to take control of large and significant areas of the French experience of the law – specifically civil and criminal trials. The first two *ordonnances* therefore represented a victory for royal power in its efforts to achieve a sought-after legal monism.

The third Louisian *ordonnance* is somewhat different both in character and in legal historical importance. This text is aimed squarely at unifying the governance of the law of commerce. This was already a delicate area because it intersected with the economic interests of powerful corporations of merchants.

I would be misrepresenting Colbert's work if I argued that his wish to regulate matters generally left to the everyday customs and practices of merchants represented nothing more than a simple affirmation of power. Instead, the complexity of this very subtle politician's aims is made fully clear if one considers the way in which the central government had the *ordonnance* drawn up. The primary responsibility was given over to professional merchants, the foremost amongst whom was one Jacques Savary, a man of little learning but who was nonetheless given the great privilege of being the authentic and experienced voice of the mercantile class. Savary came from a family of merchants and had begun trading long before he entered public service: he was, in effect, an expert practitioner.

The resulting legal document, which is relatively concise at merely 122 paragraphs, enacts only very modest technical legislation. However, it gives close attention to the merchants' desire for greater order amongst the key points of commercial law, which had developed alongside the mercantile practices of the medieval and post-medieval periods. The *ordonnance* therefore deals with such issues as commercial legal subjects (merchants, apprentices, mediators), commercial ledgers, business associations, negotiable instruments, insolvency and the commercial courts, where merchants to could have their cases heard by unrobed judges who passed reassuringly equitable rulings.

As well as signalling a victory for royal power, the third *ordonnance* represents the consolidation of an alliance that will prove extremely durable. It is no coincidence that it is the merchants who will later help the king to unify the national territory by breaking down the local legal peculiarities that favoured the feudal nobles. The third *ordonnance* may have affirmed the king's status as legislator and sovereign but it also allowed the merchants free rein (albeit with the king's blessing) to draw up their own branch of law, founded on their own customs, that was very much distinct from ordinary civil law.

The law governing matters between private citizens remained untouched by Louis XIV's *ordonnances*. The *coutumes* continued to govern these cases until the winds of revolution swept through. In the first half of the eighteenth century there were three isolated, and yet significant, reforms made by Chancellor D'Aguesseau on the subjects of gifts and of wills. However, these stick out somewhat against a background of a general absence of legislation.

By way of conclusion, we may note that, thanks to the combined labours of the jurist heirs of historicist humanism and a legislator set on asserting his sovereignty, the outlines of a national legal system, an authentic *droit français*, are becoming ever sharper.

The Kingdom of England: Constitutional Turbulence and Common Law Continuity

During the opening centuries of the modern period, France experienced a time of constitutional quiescence, with the fires of 1789's revolution left smouldering under the ever-increasing ashen weight of the Ancien Régime. Meanwhile, however, neighbouring England was in the grip of a long and tortuous constitutional struggle that would have a decisive influence not only on European constitutional history but also on the wider development of English law, including the laws governing the daily lives of citizens.

Let us start with some background for those readers who may be unfamiliar with the historical context. This period of English history is a fragmented one, with many radical shifts that scored and sometimes shattered its historical continuity. The constitutional settlement of 1215, where Magna Carta placed curbs on King John Lackland's powers, developed into that of the seventeenth century, which was still a monarchy but in a more mixed form – with an individual (the king), the aristocracy (represented by the House of Lords) and the public (represented by the House of Commons), all playing a role. By the eighteenth century this arrangement had developed in to a parliamentary monarchy, in which the government's political power stemmed from Parliament and leadership of that government was given to an individual who could command the confidence of the House of Commons. This gave rise to two new legal characters: the Cabinet and the premier.

There is certainly an element of continuity to these developments, and it lies in the progressive limitations placed on royal power, which was exercised absolutely only during the golden era of the Tudors (specifically Henry VIII and Elizabeth I). And yet such a view of events ignores the shifts provoked by the tormented political and constitutional disputes between

Parliament and monarch. The first of these shifts cannot go unmentioned and it consists in the two transformations that made the English Parliament into a novel political and legal entity: firstly, Parliament changed from a body which met occasionally and whose existence was temporary to a permanent fixture; secondly, its function changed from judicial to legislative.

Underlying this change in the superficial level of the apparatus of power of the kingdom and the struggle for supremacy that took place therein was a more hidden but, for our purposes, more interesting layer: the history of the common law. What is notable here is the stability and continuity of English law: the only real change is its ever greater infiltration into all levels of society and its enhanced prestige. It was not until Henry VIII, a monarch who initiated religious reforms in order to further his absolutist aims, that there was any attempt to discredit the common law, in favour of *civil law*, whose remote ancient Roman authorities described the sovereign as free (*absolutus*) of any limit or bond. Henry was enraged by the developments of the Middle Ages that had given the common law its own identity. There was now a substantive separation between the judiciary and the king, despite the former deriving its authority from the latter. The common law, moreover, had developed into a sort of general customary law covering the whole realm – as if it were the vessel for all the ancient customs and liberties that had accumulated over the years. The law was now seen as the vessel for a pragmatic form of reason based on an objective reading of a world of immemorial practices.

Apart from mentioning the Tudors, I would like also to clarify for the reader the relationship between common law and legislation. Here we must note the principle, laid out in the Bill of Rights of 1689, that legislative power is vested in 'the king in Parliament': a complex entity formed of the House of Commons, the House of Lords and the monarch. In formal terms, legislation most certainly had primacy over the common law. However, at least until the late nineteenth century, legislative interventions remained sporadic, unconnected and even somewhat timid. Acts of legislation were therefore naturally incorporated into a common law legal sphere that already had its own imposing shape: a system in which everything is interpreted and decided according to a centuries-old matrix of customs. Legislation always presupposed the existence of the common law and sought to integrate itself or modify what already existed. In any case, legislation and the common law were always in dialectic, making together a broad and plural legal system.

It is not misleading, and indeed may be illuminating, to draw an analogy to the relationship between the *ius commune* of medieval continental law and the laws passed by a prince or a city-state. The *ius commune* was more important to the legal fabric of a medieval political community than a city's statute or a prince's isolated edict because it was a broad and organic system of law in continuous evolution thanks to the work of jurists and the contributions of new cases to jurisprudence. In early modern English society, by comparison, the legal class drew on its robust tradition to set structuring principles which were built up into a malleable legal system, day by day, case by case. Those who wielded legislative power, meanwhile, did not attempt to challenge the common law's predominance; instead, as we have just seen, they remained somewhat timidly at the margins, not daring to meddle with a system of law that was an integral part of the English people themselves and which had become one of their defining features.

This system of law exhibits a durable continuity, something that the passage of time does not erode but rather imbues with respectability and social acceptability. Judicial rulings thus gain an undisputed primacy as a source of law.

A few of the major staging points on this expansion by continuity include: the great contribution made by the eminent judge Sir Edward Coke (1552–1634), whose minute knowledge of the medieval 'Year Books' (a sort of informal summary of legal proceedings) leads him to proclaim the supremacy of the common law courts; the security of tenure and consequent independence granted to judges by the 1701 Act of Settlement that followed the Glorious Revolution; the elevation of the status of jurists by the writings of another eminent judge, Lord Mansfield (1705–93), whose widely accepted redefinition of the role of the common lawyer changed that figure from a simple technician versed in formulae and procedures into a scholar of society with his own politics of law.

This fleeting summary of the English legal system would not be complete without a brief mention of its second type of jurisdiction, complementary to the common law: *equity*. From the fifteenth century onwards an ever greater number of legal parties were becoming disappointed with the slow pace, complicated procedures and formality of the common law courts. They would therefore petition the king's chancellor, who was normally a priest, and request that he resolve the matter following the outlook and principles of canon law. The chancellor would use the more rapid, less formal, Roman/canon law procedures reach a just resolution of the issues whilst taking note of the specific circumstances of the case. What began as a series of petitions by a few unsatisfied litigants later grew into a rival jurisdiction in its own right, in which the tradition of equity, the archetypal safety valve of canon law was uppermost.

One final addendum of note: Henry VIII's bitter schism with the Church of Rome meant that the chancellor's role was no longer filled by a priest.

Canon law was also prohibited as a legal source and removed from the university curriculum. Nonetheless, the equitable mindset lingers on and remains fundamental to the English Chancellery. The lay chancellors, all great jurists and men of culture, continue to operate within the paradigm of this necessarily pastoral form of law.

The Legal Lessons of the French Revolution

I have already referred a number of times to the French Revolution, even going so far as to drop the adjective and to refer to it as 'the revolution', as if it were the archetype of such events. I have called it the 'great revolution', in order to emphasize the scale of an event which affects not only the legal and political history of the French nation but that of the entire European continent. The revolution acts as a scythe that cuts off the old order of French society at its roots, bringing about in full a reality that Louis XIV, a century earlier, could not bring himself to countenance when he rejected the plans of his minister, Colbert. Because of this, revolution became a model for future constitutional experimentation.

The use of a singular noun, 'revolution', here risks oversimplifying the events of six years of upheavals (1789–95) which were divided into phases characterized by detectably different visions and ideologies. For example, the first phase tended to emphasize the fundamental rights of the individual within society; this was followed by a Jacobin phase that aimed to crystallize society within a rigid state apparatus that monopolized political orthodoxy and was therefore extremely controlling. Nonetheless, I am concerned here to chart the signs of disjuncture with the past that the movement brought about, the signs that show that it represented a point of irreversible fracture in the legal history of continental Europe.

The first, and most fundamental, of these signs, affecting both law and politics, is the drastic reduction of the French nation from a complex entity with societies within societies and a state that was dispersed between a variety of communitarian formations to a simple and compact structure that was rigorously, indivisibly unitary. As a legal historian, what most interests me about such a development is the means by which it is carried out, specifically the scythe of *equality*.

People and goods are viewed according to an extremely simplified model: there is only one type of legal person and that person is the equal of all others (with the exception, of course, of persons legitimately holding powers of public office). Similarly, there is only one type of legal good, which is the good precisely equal to any other type of good, and there is only one type of private property right that governs it, absolute and exclusive individual property (apart from the obvious exception of publicly significant goods). At a stroke, centuries-old stratifications and differentiations were eliminated and a radically new legal landscape was sketched out, characterized by two features that completely undid the structure of the Ancien Régime: compactness and simplicity.

Let us dwell for a moment on the result that most concerns us here: the legal equality of individuals. People became individuals who were perfectly equal in the eyes of the law, as if they were statues cast from the same mould; they became individual subjects, whose only differentiation was between men and women; they were seen as solitary subjects under the banner of one national macro-community, and no longer bound by the religious, social and professional micro-communities of the Ancien Régime, which were now abolished for good. The so-called Le Chapelier Declaration, promulgated in June 1791, which established the authority of a strand of thought that was present from outset of the revolution, called for the *anéantissement*, the absolute annihilation and elimination, of all corporate structures.[22] The people were thus identified with the nation, the people had no defining features because their innumerable and unnamed components shared none. The only function of the people was to act as mute electors of representatives, the only well-defined figures with an identifiable will, with whom the people were constrained to identify.

This very stark legal equality had the inevitable consequence of populism: the mass audience became a sort of passive platform on which the representatives played out their roles. It should be duly emphasized that the introduction of legal equality caused a total subversion of the idea of political representation: where, in the Ancien Régime, the representative was content to be the simple and innocuous bearer of the pre-existing will of individual corporations and 'estates', in the post-revolutionary period the representative himself expresses his will in a way that the mass is not able to, indeed is not even called upon to do. The rigid principle of equality deformed political representation, making it serve the so-called representative; the electoral function was therefore diminished; the body of the elected was enlarged; and a legal and political centralism was created which transformed the new state into an elite phenomenon.

This formally democratic structure could easily degenerate into a *de facto* dictatorship; and indeed this is what happened during the Jacobin period, although all the reforms were proposed as a 'dictatorship of liberty', as one of the most intransigent but also most brilliant Jacobins, Maximilien Robespierre (1758–94) put it. We should also mention the impassioned and very intelligent reflection on legal equality and political representation made by one of the protagonists of the French Enlightenment, Jacques Necker (1732–1804), in 1793, a crucial year in the revolutionary period. Necker, an unfortunate minister in the government of Louis XVI who devised reform projects that were never carried out, viewed the much-trumpeted identification between people and assembly as a construct and so as an act of deceit towards the people, despite the faith they were called on to have in their elected representatives. The very word 'representative' was a trick, according to Necker, because 'the term gives the impression of another self', an impression he saw as absolutely false.[23]

Although the process of democratization mentioned here is stifled somewhat by the Jacobin phase, we must still acknowledge the formidable novelty that lay at the heart of the legal philosophy of the revolution. This philosophy, as the reader may have noted, consisted in the demolition of derelict systems and the installation of a completely new set of informing principles, combined with a very subtle grasp of strategy. To indulge in a little phrasemaking, it was as if the surface of the French kingdom, and then the republic that followed, had been burnished to a high sheen, in which political power might see itself reflected flawlessly: any social structures barring the biological family that might compromise the compactness of the new system are eradicated, as all forms of legal particularism that might harm its unity. In other words, nation and law are in perfect unity.

Staying with this important point, the now inadmissible divisions between *droit coutumier* ('customary law') and *droit écrit* ('written law') are left behind in the shadows of the Ancien Régime, and yet it was impractical to attempt to derive a system of law from the representative body placed at the centre of the new state. The new law was to be constructed from 'principles', that is to say sweeping guidelines thought out and laid down by the new sovereign body, principles which, at the level of legal sources, would manifest themselves as 'laws' – the authoritative pronouncements of that body. The customs, of course, had to be discarded, since they originated from the ground up and from specific circumstances; jurists too, be they scholars or judges, were denied a role: partly because they had emerged diminished from the Enlightenment, but equally because they could not be controlled.

The revolution, and in this the period 1789–95 was of a piece, represents the total reduction of the law to a system of legislative acts, of *droit* to *loys*, to recall the distinction that Bodin was using as early as the end of the sixteenth century to denote this stark divide. This process now plays out in full, the wishes and the achievements of so many French monarchs and chancellors are fulfilled. Legislation becomes the only source of law since it is the only source that, thanks to its abstractness, generality and rigidity, can guarantee the legal unity of the French state. Obviously, the law is immediately provided with a supporting framework, and this soon becomes a very efficient myth-making machine.

The myths were necessary because this new type of law was an object of belief and not of critical appreciation; it was founded on axioms, that is to say on undemonstrated and undemonstrable truths. The revolution, the daughter and heir of the Enlightenment proclaimed at the outset of the 1789 *Déclaration* (para. 6) and at the end of the 1795 *Constitution* (para. 6), that 'the law is the expression of the general will' and that 'submission to the law is essential to the preservation of liberty' (*Déclaration des droits de l'homme*, 29 May 1793, para. 8), that the law 'cannot order anything but what is just and useful to society' (*Acte constitutionnel*, 24 June 1793, para. 4). Legalism and legal despotism were able to take hold because serving the abstract concept of law in general was seen as a commendable thing which did not undermine the dignity of a citizen so much as serving another man. Moreover, legal idolatry began to take hold – and would continue to dominate Europe up to the very recent past – since the law, surrounded by its impenetrable carapace of mythology, was not deserving of respect because of the justice it dispensed but simply because it was the law, and so representative of an act handed down by the supreme power.

The principles of natural law were left ever further behind, replaced instead by a statist ideology to which the legal Enlightenment had lent legitimacy with its idealization and aggrandizement of the figure of the prince. The universalizing message of the revolution was gradually reduced to a conception of the state as an entity with political and legal muscle.

At the heart of this revolutionary legalism lay a specific undertaking, on which the *Constitution française* of 1791 ends, and which bears rereading: 'A code of civil law common to the whole Realm shall be laid out' (II, Fundamental Provisions, final paragraph). The revolutionary constitution does not merely deal with this code in generic terms; rather, it sets out the task in detail: the civil law that the *ordonnances* of the Ancien Régime had left to the age-old tutelage of custom is to be brought under the aegis of legislation. Legislation would now be the hallmark of innovation; the whole legal order was to be legislated for.

The French Revolution was a bourgeois one, carried out by the third 'estate', the city-dwelling classes, and the revolutionaries moved to take control of those things that were particularly dear to them, such as property and contracts. They drew up for these areas a framework that chimed with their interests. The central aspects of private law, left untouched since the reforms of commerce in 1673, were finally to be regulated.

The Age of the Code

From the first moments of their movement, therefore, the revolutionaries fixed their sights upon a code (the first proposals date from 1790). They desired a code, indeed they demanded that the area of law yet to be codified, civil law, be brought under the legislative banner. The revolution thus provides a faithful reflection of the age of the code, an age of which it was both offspring and progenitor. What should this mean? And, above all, why the singular 'code' in the heading to this section and in its opening paragraph?

The singular is intentional because it serves to distinguish between the code discussed here and the many codes that abound in antiquity and early modernity, since, as we have seen, the Latin noun *codex* or the French *code*, to give two examples, are often used to designate collections and compilations of laws made on the initiative of individual jurists or monarchs. One Italian legal historian has gone so far as to suggest calling such collections and compilations *consolidations*, in order to separate them from *codifications* and to allow the post-Enlightenment project of the Code to be understood in its full historical specificity.[24]

The number of the noun, as I have already pointed out, should not be misconstrued. The code that takes shape as the newest legal source in the late eighteenth and early nineteenth centuries, and that will dominate all of the latter century and much of the twentieth is more of an idea – a political and legal project. It is a project that seeks to propose a radical solution to the problem of the sources of law. This was a problem which the legal Enlightenment had so lucidly divined, with which certain sovereigns sensitive to the new thinking had wrestled without even realizing, which the revolutionary scythe had finally swept aside because of the radical nature of its break with the past, and which Napoleon, in this case a true heir of the revolutionary message, would finally resolve with his far-reaching and comprehensive act of codification.

The prerequisites of the code were two revolutions: the cultural revolution of natural law and the Enlightenment and the historical events of 1789–95. These led to two fundamental innovations: a new way of understanding the relationship between political power and the legal sphere and, as a consequence, a new way of conceiving of and carrying out the production of law.

For the whole of the medieval period, and lasting well into the Ancien Régime, the production of law displayed three definable characteristics: it was incremental – legal sources tended to sediment themselves upon others, causing confusions and uncertainties; it was pluralist – the law was derived from a number of sources even if early modernity had tended to define the prince-legislator as an ever more imposing figure; it was separate from the state – although there was an increase in the number of legislative acts, the endurance of custom-based civil law confirms this. At root, all of these characteristics grew out of a vivid appreciation of the historicity of the law, that is to say of its allegiance to the ever-changing nature of human society, rather than to the calcifying apparatus of the state.

However, the intellectual revolution wrought by the doctrine of natural law and by the Enlightenment, which reasserted the possibility of deciphering the perfect forms of the natural world, brought with it a horror of the sedimentary nature of the old legal order, whose historicity was now seen as disorganized and muddled complexity. This was a great innovating project that gave rise to the rediscovery of the state of nature and of the authentic state of mankind before the ruinous intervention of history, and faithfully granted the prince-legislator and his supreme representative, the law, the power to transform the chaos of the previous system of law into geometrically clear and certain legal principles. But it could only have one result for the production of law: a complete take-over of the legal order by the state. Political power inevitably became the supreme and only source of law; a rigorous legal monism replaced the earlier pluralism. What I have defined above as *legal absolutism* now conditioned the shape of the law.

The code is a faithful and prototypical expression of this legal absolutism. It was, and I repeat this so the reader may bear it in mind when reading about it, a completely new way of conceiving of and carrying out the production of law. Three fundamental tensions underlie the code and shape its development: it must be a unified source of law, a mirror for the state and the mortar that holds its bricks together; it must, as far as possible, be an all-embracing source in order to guarantee the desired unity; and, to the same end, it must be the exclusive source of law.

One must not forget that the code is the final product of a general attitude of legislative mysticism that turned the law into a myth and a cult. The code must therefore be seen in the context of a now entirely monist mindset that placed the written law at the head of an extremely rigid hierarchy, above all other sources of law. Nor should we neglect the code's structure, which seeks to encompass all human experience in a comprehensive and detailed series of written rules that consider, define and govern all social institutions and their possible realizations with studied precision.

The thought processes of natural law also impose a radical innovation on the content of the code: behind the general attitude towards historical specificity that sees it as a morass of contingent accretions that obscure genuine naturalness, the human subjects and relationships treated are abstract. The result is that the code's reality is populated by virtual beings – as virtual as the natural state that is taken as the code's exemplar. These virtual beings are models of men rather than flesh-and-blood entities: the result is that the code and its purist formulations look beyond the historical moment that created them towards a timeless horizon.

I should make one thing clear: certain legal historians, who see the code as the final link in a long chain rather than as the first link in a new chain, as I have argued here, link the late eighteenth-century code to the many experimental attempts in the near and recent past to organize and systematize the sources of law. Such scholars make a great deal of the continuing presence of a variety of structures from the legal system of the Ancien Régime in documents such as the *Code Napoléon* of 1804. These presences are undeniable, and I would take them for granted given that common sense and good methodology both dictate that no historical phenomenon is born *ex nihilo*; it is therefore not viable to argue that the umbilical cord linking present to past would be completely severed in this instance. Once could add that the drafters on the Napoleonic commissions, to continue with the same example, were inevitably individuals born and brought up under the Ancien Régime and so necessarily had some ties to its solutions, at least unconsciously.

But these continuities should not cause us to fall into the trap of giving too much weight to specific parts of the code that do not necessarily fit well with the greater whole. The most important aspect of the code is the self-important mindset of rejection of the past, the radical choices made in the matter of the production of laws and the radical reform that puts an entirely new face on the system of legal sources.

To that we should add that, in the matter of content as well, innovations are apparent at every turn: there are no more classes, associations or institutions – only an individual reflective of an active subject in the state of nature. This subject is universal: there are no noble subjects or plebeian subjects, no peasant or merchant subjects, not even rich or poor subjects (the poor have been abolished in this utopian world of models!). Another sweeping innovation is, as we shall soon see, the tools put at the disposal of the code: private property and contractual agreements – a now reunified concept of property and a form of contract that expresses only the unfettered consent of its free participants.

To bring this story to a close, it seems that Napoleon was the first comprehensive codifier, with the possible exception of an experiment conducted in 1787 by the Austrian emperor Joseph II but confined to the area of criminal law – a very nationalistic field highly constrained by the exercise of sovereignty.

A well-informed reader might well protest at such a bold statement, pointing to a great legislative project from the kingdom of Prussia, promulgated in 1794: the *Allgemeines Landrecht für di Preussischen Staaten* (the 'General State Laws for the Prussian States', often known as the ALR). De Tocqueville, the well-known and perceptive French intellectual, was rather disdainful of the project, opining that 'below the completely modern head, one can see a very Gothic body taking shape'.[25]

De Tocqueville was quite right. The idea for legislation came at the behest of the Enlightenment king Frederick II, but over its fifty years of gestation it had suffered so many attacks and actions of resistance from the conservative classes of the realm that it was quite transformed by the time it arrived at its rather dispiriting result: a sprawling, elephantine mass of rules quite in contravention of the Enlightenment ideals of simplicity and clarity and, moreover, centred around the unchecked continuation of class (*Stand*) as the protagonist of society – a further betrayal of Enlightenment individualism. But the two determining features, which prevent me from allowing that the *Landrecht* may be defined as a code, are that not only was it not an exclusive body of law, since it left local laws in place and sought to replace only the old common, jurist-generated law, but moreover that it defined itself as a mere subsidiary to the local law.

French Codification of the Early Nineteenth Century in Detail

The nineteenth century appears to a legal historian's eye as the age of the code in its fullest manifestation and maturity of expression. The century begins with two great acts of codification, the French and the Austrian (of 1804 and 1811 respectively), whilst the promulgation and later entry into force of another equally significant codification, that of Germany (1896–1900), marks the move to the next hundred years. The intervening period is studded with codes, all of them *codes* in the narrower, legal historical, definition of this ambiguous term I have set out above.

Although the trend towards codification characterizes almost all branches of the law, I shall be most concerned here with the experiments in the arena of civil law. This is for two fundamental reasons: firstly because, as I said at the outset, my vision of the law in this book is as the expression of a communal mentality and therefore it is in relationships between individuals rather than in the rules laid down by the centres of power that we can best observe it at work; secondly because, in a bourgeois society lacking a written constitution, it is the civil code that provides that society with its true constitutional structure. A first demonstration of this can be found in the not insignificant fact that both the French and the Austrian codes, which I am about to discuss, open with a number of provisions relating to more general legal issues (six articles in the French code and fourteen paragraphs in the Austrian). These preambles are a sign of the particular importance that the two legislators attributed to these new social orderings.

There is another factor that unites these two codes that I should note before describing each one in detail: despite the fact that they were promulgated two centuries ago, each one is still in force in its respective nation – the French Republic and the Federal Austrian Republic. There have, of course, been many adjustments in that time to bring the codes up to date; nonetheless, it is a fact that underlines their significance without need for further comment.

The French Civil Code, promulgated in 1804 and universally known as the *Code Napoléon*, has very little in common with previous legislative consolidations because it is a bona fide product of the legal Enlightenment and of the French Revolution. The Napoleonic Code is a completely new species of legal source, a new means of producing law; it does not aim to perfect the legal structures of the past, not even those of the recent past, but instead sets up new rules that look towards the future, and even towards eternity. As such, the code is not a remade ensemble of old rules newly selected and ordered but an organic whole, underpinned by a unitary project, hedged by rigorous guidelines, thoroughly coherent and logically structured.

The code provides, in effect, clear, simple and abstract law as foreseen by the reductionism of the Enlightenment. Two completely unitary notions of person and good stand out, made unitary by the scythe of equality that has completely cut away all the former social and economic complexities. The Napoleonic Code presents the person and the good in their natural state; it liberates them from any historical superstructure, and rediscovers them in their so-called primal originality.

Regarding the person: the onus of equality does not tolerate any diversity that might ensue from membership of a religious faith (the law's point of view is – for the first time – completely non-confessional), nor from the participation in any of the corrupting old intermediary social structures. Persons are now abstract, socially naked individuals.

Napoleon was by no means wedded to the ideals of the revolution, nor was he moved by the sincere passion for social reform that hummed through the Jacobin era, but he nonetheless continued and even intensified the revolutionary and Enlightenment idea of the law as an indispensable check on society and the necessary foundation of power. The process of codification fits with Napoleon's despotic grip on the French state: Napoleon the *condottiere*, the future emperor, would take care of proceedings himself. Bonaparte divined the enormous significance of bringing private law under the aegis of legislation. He thus began his work as a legislator with the area of civil law, which up to a few years before had been off limits to legislation, attending the sittings of the drafting committee assiduously and frequently imposing his own ideas.

The result was extraordinary: a coherent document comprising 2,281 articles in which all conceivable interactions between private citizens were enmeshed in an meticulous system of principles, definitions, provisions, obligations and sanctions. Although among the very jurists who drafted the document (of whom we should at least mention Jean-Étienne-Marie Portalis) there was an awareness that such all-encompassing coverage was an unattainable mirage, nonetheless their efforts were directed towards constructing a framework of rules that was, as far as possible, totalizing. *All* conceivable civil law was encompassed by the compact walls of the Napoleonic Code. This compactness was reinforced by the total eradication of customary law, practical jurisprudence and legal scholarship, all now banished from the fold of sources of law by the simple will of a legislator. And Article 4, which bars judges from refusing to reach judgment 'on the pretext of the silence, obscurity or defective drafting of the law' even in preparatory contexts, may have seemed a valid means of empowering judges to one or other of the drafters, but, in the wider context of the code, it soon became the bolt that shut the judge in a cage of rigid legalism.

Legal absolutism triumphed with the Civil Code, wherein it reached the zenith of its intensity.

The nature of the substantive content treated by the code was also completely new, although old threads were certainly discernible in the new fabric (and it is the fabric, after all, that counts). The code was divided into three sections: the first dedicated to persons, the second to property and the third to means by which property might circulate. The cornerstone of the code is private property – even the persons treated in the first section are discussed mainly from the standpoint of the property they own, whilst the third section deals with instruments that exist for the purpose of allowing goods to circulate (contracts, liabilities, successions, donations).

Property is no longer a combination of powers over an object as it was from the Middle Ages until the French Revolution; now it is a unitary phenomenon bound tightly to the freedom of the individual subject and, therefore, the single and indivisible influence of a person over a thing – an influence that is therefore absolute, perpetual and exclusive. The odd trace of the past can be detected nonetheless, although these traces sound distinctly out of tune within the new structure: the title of the second section, 'Of Goods and of the Various Modifications of Property', makes real but limited rights sound like mere 'modifications' and points towards a concept of property as a complex and composite reality. Similarly, Article 544 attempts a definition of property and yet contains a manifest contradiction between its affirmation of the absolute nature of property owners' rights and its granting to them of no further powers than 'use' and 'disposal'.

The area of contract law is also significant: all of the types of contract that have specific uses in daily life and the practice of business are of course considered (sale, exchange, rental, loan, etc.), but before and above these there is the *contract*: the expression of free consent that, since it is an expression of individual liberty, has legal consequences and is in and of itself worthy of safeguarding. The private citizen is allowed a wide compass whose parameters are set by the legislator, within which he may create such agreements as may please or serve. Articles 1101 to 1167 create in the middle of the code a sort of general theory of contract law. Here are found treated issues such as consent, legal capacity, contractual subject matter and suits – all the prerequisites not for a specific type of contract but for any free interaction between the desires of two or more people who govern their own lives.

The contract, as described in Article 1134 – and the description is significant for such a legalistic culture – is a piece of law that binds the parties. This is the law of the realm inhabited by the French citizen, particularly the businessman. The solid alliance between political and economic power, which we saw already in evidence in the *ordonnance* of 1673, finds full confirmation here in the Napoleonic Code, and in the Commercial Code which followed, only a few years later, in 1807.

Austrian Codification of the Nineteenth Century in Detail

The Napoleonic Code was able to be so organic and coherent because it was put together immediately after the years of revolutionary upheaval which tore down the old encumbering legal structures. The same cannot be said for the other great code that opens the nineteenth century, the General Civil Code, or *Allgemeines Bürgerliches Gesetzbuch* (often referred to by its abbreviation, ABGB), which was promulgated in his hereditary German lands in 1811 by Francis I, the last Holy Roman Emperor and the first emperor of Austria. (The code was subsequently extended to the Viceroyalty of Lombardy and Venetia on 1 January 1816.) In Vienna, unlike in Paris, there was still a manifest contradiction between the central apparatus of government, in which the Enlightenment ideas of the eighteenth-century reformist monarchs were circulating, and the wider socio-economic structure that remained solidly class-based and feudal. Past and present coexisted antinomically, and the new code reflects that contradiction faithfully.

There is no doubt that the ABGB is a *code*, given the essential choices that characterize it. The Austrian code bears witness to a new conception of the production of law which views a code as necessary. Given this, the ABGB is a new source of law, which asserts its monopoly over lawmaking by abolishing the common law and negating the autonomy of traditions and local statutes (paras. 10 and 11). It is new also because its legal subject is assumed to be the unitary subject of natural law with all his inalienable rights (para. 16). In effect, the ABGB is shot through with an individualistic vision of the law taken from the philosophy of Immanuel Kant. The man largely responsible for this Kantian influence was Franz von Zeiller (1751–1828), a high-ranking civil servant and magistrate, and the jurist who made the greatest contribution to the shape of the ABGB's final version.

The ABGB is particular also in the drafting technique that it adopts. The Austrian drafters are careful to maintain certainty and clarity throughout the document, using a plain and unadorned language free of any complex jargon. They do not chase after nuances or minutiae, with the result that the code becomes an organic exposition of principles. Indeed the 1502 paragraphs of the Austrian code seem to add up to a considerably more measured legal document than those of its French counterpart.

The ABGB's architecture is central to the project: after a brief introduction, the work is divided into three parts, the first devoted to the 'law of persons', the second to the 'law of things' and the third to the 'provisions common to the laws of people and the laws of things'.

Within the introduction, paragraph 7 especially stands out, since it confronts the main problem faced by any codifier: the issue of potential lacunae left in the system set up by the code. The solution is a daring one which breathes life into the stagnant paragraphs of codified material: if literal and logical interpretations do not suffice, and if, even after reference to 'similar cases decided more precisely in law and to the purposes of other similar laws, the case still remains doubtful, it shall be decided according to the principles of natural justice, having regard to the circumstances as can most diligently be established and most carefully considered'.[26] The phrasing here clearly demonstrates how the two great forces of the previous phase of Austrian legal history, natural law and common law, were still uppermost in the minds of the codifiers. Natural law had been fiercely championed by many eighteenth-century sovereigns and ministers at the Viennese court; the common law, meanwhile, had not suffered the same defeats in Austria as it had in France and was still vigorous enough to gain a place amongst the paragraphs of the code where reference was made to particular circumstances that only a judge or other interpreter could gather together and evaluate.

But it is in the content of the code that one can most easily observe the intimate incoherence between the document's extremely innovative structure and the moribund socio-economic fabric that was unthinkingly carried over into the Austrian empire.

There is no exclusion of intermediary social structures, which are even explicitly referred to at paragraph 26. With regard to contracts there is a contradiction between the general theory of contracts laid out from paragraph 859 onwards, which privileges the creative power of free consent, and a framework for contracts of sale that still reflects the system prevalent in Roman law and the common law. Under these systems, simple consent is not held to be sufficient to enact the transfer of a good from one person's ownership to another; instead the change of hands (*traditio*) is held to be essential to the change in owners.

However, it is in the area of royal prerogatives that the relics of the old ways remain most obstinately present, muddying the supposed transparency of natural law. Property is by no means the new unitary phenomenon of the Napoleonic Code; the ABGB's palette of ideas and terms is still that of the glossators and commentators who conceived of property as a composite reality, an aggregation of powers – terms such as 'demesne' and 'tenancy' are still used without a second thought. The ABGB also continues with a number of structures that were strongly opposed by the forces of the legal Enlightenment and so were regarded by the revolutionaries as relics of the past that were therefore unacceptable in a society based on new foundations. These structures included: the fee simple (perpetual lease), which fragmented the concept of property by making permanent the tenant's right to enjoyment; fee tails (hereditary leases), which impeded the free exchange of property; trusts, which also limited the free exchange of inherited property since they might bind the beneficiary to conserve and transmit the legacy to a third person, or stipulate the property's inalienability. Paragraph 618 and following, together with paragraphs 1122 and 1124, give full authority to these old structures that had fitted so well with the socio-economic makeup of medieval and post-medieval societies.

These structures are not barely visible threads that disappear amongst the breadth of the fabric; such relics, like the ones we saw when studying the Napoleonic Code, are often due to the drafters' legal education under the Ancien Régime and so they can be counted as a more or less unconscious throwback if one takes a proper historical view of the law. In the ABGB, however, we find an overall vision of property rights that is in conflict with the new vision of the legal subject; the older structures are transplanted whole and constitute foreign bodies inside the larger organism of the code. The Austrian code, in effect, shows two different faces to us.

It is this ambivalence that means that, unlike the Napoleonic Code, the ABGB will have only a limited influence on the legislative developments of the nineteenth century.

Law, Legal Scholarship and Legislative Practice in the Age of the Code

To the nineteenth-century bourgeoisie, the code appeared as a great legislative triumph, perhaps the last word in human progress. The code was the full incarnation of all the ideals of the Enlightenment: it was a simple, clear, certain, universal, and therefore egalitarian, law. The protean system of the early modern period had been reduced to harmony and coherence, fulfilling the eighteenth-century desire for geometrical order. The law had been fixed in its entirety within one text, which any citizen could read and understand. The law had been legislated: it was under the aegis of the political structures controlled by the bourgeoisie, who were now able to view the code with satisfaction as the fruit of all their effective manoeuvrings.

The code was the fulcrum of the bourgeois legal order, it was its reassuring written constitution, free from the numerous idiosyncrasies of customary law, the strange contributions of unhinged jurists, the worrying decisions of judges. It was the cornerstone of the new legal edifice, that would endure, that would deserve to endure and that would deserve to be spread.

And so for the whole century there reigned a mania for codification. In France, the first decade was dedicated to completing the process of codification. After the first and most notable product, the Napoleonic Code, came the Code of Civil Procedure, the Commercial Code, the Code of Penal Procedure, the Penal Code. Each of these efforts functioned as a model to be followed, a model that was followed because of its lack of ambiguity, its linear and coherent trajectory. Codification was thus easily pushed through in other European societies in which the bourgeoisie was similarly powerful.

The French model was followed in many of the statelets of pre-unification Italy at the time of the European Restoration after the Council of Vienna of 1814, even though most of these nations modelled their political structures on those of Vienna. The civil codes passed in the Netherlands in 1838, in Portugal in 1867 and in Spain in 1889 all bear the Napoleonic stamp. The same can be said for the results of the first legal codification in post-Risorgimento Italy in 1865: many of the articles of the Italian codes are in fact word-for-word translations of their French counterparts. The Italian Civil Code, however, does show some significant signs of autonomy: for example, in the 'Preliminary Dispositions to the Civil Code', whose third article treats the issues of lacunae and posits the recourse to 'general principles of the law', a generic formulation that allows the judge at least some leeway to look beyond the system of positive laws. Article 2 of the first volume also refers to 'moral bodies', or legal persons, an idea completely foreign to the Napoleonic Code but retained here, albeit in a somewhat diffident insertion in the realm of public law.[27]

The above-mentioned Spanish legal code of 1889 also needs some further treatment, given Spain's very particular legal history. Spain was already a unified nation in the fifteenth century, but until the nineteenth there remained in force a system of *legal localism*. This began to be reformed under the rule of Joseph Bonaparte, and was entwined with (and complicated by) what was called *la cuestión civil foral*, or the need to respect the systems of civil law of other parts of Spain. *Foral* is an adjective derived from the noun *fuero*, a central term in Spanish law since the early Middle Ages, and one which designates a local customary legal system that grows up over the course of time and comes to characterize the historical identity of a given *nation*. In the nineteenth century five legal traditions belonging to the five nations of Castile, Aragon, Catalonia, Navarre and Biscay were strongly distinguished and we should add to these the clearly localized systems of Galicia and the Balearic Islands. The *cuestión civil foral* was the greatest and most intractable problem facing the many codificatory projects that dot Spain's nineteenth-century history. The result was that the code finally promulgated in 1889, despite being based in outline on the Napoleonic Code of 1804 and the Italian Code of 1865, was cast as the common law of the realm but only supplementary to the *derechos civiles forales* ('local civil law') which were to be drafted as 'appendices' to the code (art. 12, para. 2). The term *appendix* was to be understood in a restrictive sense, so as not to undermine the effort of codification, so painstakingly completed.

Codification continued throughout the century because politicians and jurists were convinced that they had found in the code the remedy to all the evils that had previously afflicted the law. This persuasion was closely linked to their certainty of having constructed a stable society founded on shared values. The code, as the litany of those values, could easily be attributed with the fixity of general and eternal rules – it may as well have been carved in stone as written on paper. This was of course naive optimism, which had failed to take into account two factors which would erode the code's stability: technological and economic development, and social change, especially the increasing importance of the working class.

During the first years of the nineteenth century, immediately after the revolutionary uprising, the civil code represented the fulfilment of the historic aspirations of the bourgeoisie to own land; the land was now freed from the stranglehold of class-based constraints and property charges and there was now a framework for its free acquisition. This had very specific results in the economic sphere: the code's model of property was a rural one and so the concept of real property was paramount. But, in the decades after codification there was great technological progress and, as the French economy became more dynamic, industry began to play a pivotal role. A new type of wealth began to dominate economic commerce, relegating real property transactions to the second rank of importance. As early in the 1830s Pellegrino Rossi, an Italian intellectual equally well versed in law and economics, lucidly contemplated contemporary developments and was moved to draw some dispiriting conclusions in a speech to the Academy of Moral Sciences in Paris. Rossi remarked on the lack of economic knowledge amongst Napoleon's codifiers and the inadequacy of the code to match up to the new and increasing demands placed upon it. The book that was to be set in stone had been rapidly overtaken by events.[28]

There is a further, even more prejudicial, factor to be taken into account. The code established social order by offering the lentil broth of legal equality to the masses whilst leaving the inequalities of fact untouched. Such an arrangement could only continue as long as the proletariat did not attain class consciousness and begin to foment protest and revolt. As soon as the bean-counting, classist bourgeois state began to come under attack through workers' demands, strikes and demonstrations, the code would reveal the impotence of its basis in the generalities and abstractions of natural law. The meta-history of natural law's founding myths would have to give way to the gritty, unforgiving, flesh-and-blood facts of daily life.

And so, in the second half of the nineteenth century, the magical virtues of the code became somewhat obscured, and special legislative acts began to be introduced to supplement the august mother of laws. These acts did not counteract the code, but they added to it specific provisions aimed at taking into account, albeit often reluctantly, the profound social, economic and technological changes that had taken place. Nonetheless, governments tried to legislate only when they had to, seeking to keep the special laws to a minimum. As certain Italian civil lawyers put it, the new laws were like satellites orbiting the main body of the civil code, which continued to dominate the legal order of the nineteenth-century state.

To sum up: the nineteenth century was the age of codes, the civil code to be precise, but as time passed, the manifestations of the law became more complex as the bourgeois legislators were forced to pass a series of special laws in order to respond to new legal needs.

There is one question that needs to be answered at this point, and it is implicit in the title of this section: what space is allowed to jurists, both theoretical and practical, in the midst of the general legalistic sclerosis that dominates great powers of Europe and their colonies (or ex-colonies)? The answer is simple: if one takes the poisonous polemics of the Enlightenment thinkers against jurists, together with the atmosphere of legalism which began with the French Revolution, together with the legal monopoly instituted by Napoleon with the code, there is very little space for such figures. And so, the term customarily used to describe the work of jurists in the wake of the codes is the expressive word *exegesis*.

Exegesis is a term handed down from the ancient practice of commentating on the Holy Scriptures. It implies that the scholar should adopt a rigorously passive attitude to a text that is held to be of divine provenance and aim for a purely expository function with regard to the text's intention. The transfer of this sacred word to a secular sphere demonstrates the extent to which the legislator's will had become exalted. The intellectual labour of an interpreter of the law such as a scholar or a judge is thus reduced to pure knowledge, to an exercise of logic: he may be permitted to bring together scattered fragments of positive law but is constantly enjoined not to trespass the boundaries set by the legislated law.

Certain civil lawyers and legal historians have tried of late to begin a re-evaluation of the dense forest of literature, mainly consisting of lengthy commentaries, that grew up particularly in France, Belgium and Italy after the passing of the codes. In my view, this is a futile effort. It is true that the writers are no sheep and that the more thoughtful exponents of exegetical literature certainly produced works of admirable textual intelligence, clarity and logical rigour. However, it is equally clear that their works are suffused with the legal idolatry of the times and adopt a servile mentality with regard to the law's imposing presence. Despite the fact that the jurists were the first victims of the legal sclerosis of the era, since they were deprived of their traditional role as sources of law, they were nonetheless thoroughly proud of presenting themselves as *servi legum* ('servants of the law'), in awe of the indisputable majesty of that supreme product of human progress – the code.

This timorous state of mind is visible in the lack of any contribution by legal scholars to meeting the new demands and problems of a changed society. Jurists were unable to lift their eyes from their legal texts in order to face the troubles that were boiling up in society and seek to solve them. The text had primacy and respect for it remained the jurist's defining attribute. This passivity was what led, at the end of the nineteenth century, writers including Saleilles and Gény to raise forceful and courageous calls for a breach to be made in the now encircling walls of the code.

Practical jurisprudence presents a different picture: since the discipline was by its nature in the front line of society, it could not avoid contact with the white heat of social, economic and technological development. In the contrast between old law and new facts the most sensitive judges rediscovered their mediatory role and behaved in a not dissimilar manner to the medieval jurists with their ancient Roman texts. But of that more below.

The Law, Legal Scholarship and Legal Practice in the German-Speaking Region during the Early Modern Period

We have touched only briefly thus far on German-speaking areas: a mention in the section dedicated to the Middle Ages in order to note the absolute predominance of customs there; and, in our consideration of the modern period, nods to the Lutheran protests and the emergence of a strong state apparatus in Prussia in the eighteenth century. The project of this book, which seeks to outline the essentials of the development of the law in the European continent, has necessitated the knowing omission of the lively currents of legal humanism that coursed through the sixteenth-century Germanic cities of the upper Rhine and of the valuable contributions of German legal philosophy to the cultural revolution of the Enlightenment in the eighteenth century.

However, our wandering path now brings us to a more extended stay amongst the Germanic people and to a more detailed analysis of their doings, since we are now treating nineteenth-century Europe, where German-speaking areas become pivotal and indispensable to any historical account of the period. It would be difficult to understand the complex and particular legal situation of Germanic high modernity if we did not follow its story back somewhat. But this turning back on ourselves is no bad thing, since it will allow us to fill in the lacunae mentioned above. I may somewhat sacrifice a truly nuanced account of the trajectory we trace here, but this is a price I have already accepted in order to be able to furnish the reader with an account of the essential developments in the field of European legal history.

In the Middle Ages and up to the thirteenth century, the political sphere of the Germanic area was characterized by an extremely fragmented structure based on feudalism. As I noted above, its accompanying legal sphere was dominated by customs that, in their crude immediacy, faithfully reflected the local collective consciousness. Isolated from the great waves of scholarly reflection that were already sweeping the Mediterranean regions, this rich but unrefined system of customs was efficiently interpreted by judges. The Germanic judges lacked any capability to theorize because they were not men of culture but were instead experts in customary practices; they were granted a great deal of social prestige thanks to their universal membership of the noble classes.

It was undoubtedly a rudimentary legal system, but it was able to function satisfactorily in a static and predominantly agrarian society. However, by the time of the late Middle Ages, when economic and political developments meant that this simple society was becoming more dynamic and complex, customs and practical judges no longer sufficed. At this point a historical process which would define future experiences of the law in Germanic lands began; this was known by the term *Rezeption* ('reception'). It is a familiar term, but one that might prove misleading if it were understood as indicated the acceptance and incorporation of a foreign body of law into the living Germanic legal system. I shall seek to clarify its substantial historical significance here by boiling down its multifaceted development as much as possible. Historians of German law tend to see *Rezeption* as taking place in a number of phases, each characterized by its own typifying features.

And so, what happened in Germanic legal history between the thirteenth and the sixteenth centuries? During this time, society underwent certain profound changes, especially on the economic front. The economy became more diversified and so needed a more national structure, which in turn necessitated general theoretical models and organizing categories. Economic development therefore demanded a legal culture that possessed an arsenal of principles and concepts capable of combating the outbursts of the real world of facts. What we usually call *Rezeption* is thus, in reality, a response to the demands that the changed social realities of German society had begun to make. These developments engendered a thirst for legal learning, and the study of law in German-speaking lands took on a more scholarly bent, something that had previously not been the case but which now came to seem inevitable. This new legal scholasticism was necessarily accompanied by the arrival of people and ideas from the south of Europe, who brought with them the rich assortment of theoretical approaches developed by the hordes of glossators and commentators on Justinian's *Corpus iuris*.

New social demands, and the arrival of new people and of a new tradition of legal culture, all contributed to the creation of a new legal order. The impeccably scholarly *ius commune*, taught by professors at the universities of Italy, France and Spain, institutions that not only already existed but which already possessed no little prestige, was the foundation of the much-needed new German legal system. At first, the Church provided the most suitable conduit through which the new law might arrive, primarily because of its judicial role and the army of jurists it possessed. But soon the trickle of clergy became a flood of all types of lawyer, transforming late medieval Germany into a true melting-pot. German students flocked to Italian universities – especially Bologna and Padua – and brought back their acquired legal wisdom to their homeland. Meanwhile, the universities founded in German-speaking lands during the fourteenth century, which taught canon law at first and then the *ius commune* thereafter, began to make substantial strides towards the creation of a community of jurists that was radically new in its educational composition and its specific expertise.

This peaceful conquest not only installed people with novel technical and cultural backgrounds to university chairs, its influence reached as far as the imperial court, city governments and every level of bureaucracy and judicial administration. In practical terms it was a decisive conquest which enacted a wholesale transformation of German law. Thanks to the technical and conceptual implements brought from the south of Europe, the law became much better able to shepherd in the socio-economic changes in due course. In 1495 the Diet of Worms, a deliberative assembly of prelates and noblemen convened at the urging of the princes of the Holy Roman Empire, persuaded Emperor Maximilian to sign an order which re-established an overarching judicial institution in the empire separate from the emperor's personal influence: the Supreme Court (Reichskammergericht).

This apex of the Germanic legal system sat in the free city of Frankfurt am Main. It is very significant for German legal history that eight of the sixteen justices were required to be trained in common law and that the court was required to decide cases according to common law as the legal framework for the whole of the empire, unless one of the parties specifically invoked their local system. The *ius commune*, that is the canon and Roman law as reshaped by medieval jurists, represented the now reaffirmed legal unity of the Holy Roman Empire, despite its remaining politically extremely fragmentary and so essentially impotent.

This transformative process, traditionally known as the *Rezeption*, had a profound effect on the legal history of the German-speaking lands in the early modern period. Whilst the political unity of the kingdom of France led slowly but surely to the centrality of the sovereign's will and the preponderance of legislation as a means of making law, in Germany the protagonists of the legal system continued to be jurists, theoreticians and practitioners, university professors and judges.

Indeed, between the sixteenth and the nineteenth centuries, we may observe a community of technical adepts who, rather than seeking to raise up towers of theory that were walled off from day-to-day concerns, threw themselves into their role of creating social order. The jurists tried to interpret the legal consciousness of the German people, creating a system of case law (that is a jurisprudence that draws its reasoning from concrete cases), which was tolerant of such local law and indigenous customs that the *Rezeption* had left untouched. The period between the sixteenth and the nineteenth centuries in

Germany bore witness to a flourishing of law that is usually termed the *usus modernus Pandectarum* – the modern revival of the old Justinian *Corpus iuris* as interpreted by medieval scholars. This development is sometimes called the *mores hodierni* ('customs of today') or the *nova practica* ('new method') – terms which reveal the character of the changes as scholarship deeply rooted in daily legal practice. Indeed, so grounded was the new system that it soon became blended with the local legal customs and judicial procedures. Medieval common law had adapted Roman law for a new set of historical circumstances; we see the same confident reinterpretation in sixteenth- to nineteenth-century Germany. Scholarship and legal practice combined harmoniously, and an authentically German form of legal learning which claimed and aspired to express the legal consciousness of the German-speaking Reich was born.

Alongside this widespread attitude, from the early seventeenth century onwards there were reforms in the teaching arrangements of Germanic universities that were targeted at giving the students a preponderantly practical training. The professors were not called on to be faithful Roman lawyers so much as individuals who adapted the ancient Roman texts *ad praesentis saeculi usum* ('to modern-day practice'), just as the young students would be expected to do in their work as judges, notaries and lawyers.

Law, Legal Scholarship and Legal Practice in the German-Speaking Region during the Nineteenth Century. The Historicist School of Law and Pandectism. The Construction of the Legal System

The preceding section, which may well have seemed like a digression to the reader given its concentration on late medieval Germany, aimed to explain the distant historical reasons behind the singular situation of the law in Germany at the outset of the nineteenth century. On the one hand there reigned the geometric constructions of natural law, with its exaggerated abstractions and its rejection of any grounding in concrete experience; on the other there persisted this *usus modernus Pandectarum*, which adulterated ancient Roman law by mixing it unconcernedly with local German sources in the name of legal pragmatism.

The end of the eighteenth century and the beginning of the nineteenth in Germany sees the formation of the so-called historicist school of law, which sets itself against both of these two dominant traditions, seeing each of them as a reductive approach to the law. The historicist school takes its methodological foundations and typifying features from the innovative views of that most influential current of thought: German Romanticism. Romanticism entailed a scepticism towards the Enlightenment conception of progress, a horror of the idea of confining society in purely rational, geometric models, a respect for the past and a love of tradition, and a re-evaluation of history in all its complexities – paying special attention to the irrational dimensions of religion, belief, custom and unconscious conventions. The new Romantic consciousness seizes on these irrational aspects of history and makes them the foundations for a new legal edifice, whose defining characteristic is located not in displays of power, in laws or in codes, but rather in the day-to-day life of the people.

From a Romantic point of view, custom gave the law its best expression, since it was a malleable entity generated from the lower reaches of the community and, as such, in harmony with the dynamic nature of society. The great rediscovery of the Romantic age was the historicity of the law, something that could be demonstrated by the oft-repeated analogy between the law and language, which could be two very effective ordering principles of the social sphere in which subjects moved so long as they remained agile enough to keep pace with the progress of the underlying social structures.

Given this very diverse profile of legal culture, in which natural law, *usus modernus* and the historicist school all competed with one another, the first problem that emerged after Germany had been struck by the Napoleonic hurricane was that of which legal sources were suitable for the needs of German society in the contemporary context. This question was rendered particularly troubling given the inconvenient presence of the French codifying model. More inconvenient still was the adoption of a code by the Austrian empire in 1811, not to mention the expansion of the Napoleonic empire into German-speaking territories, bringing with it the Napoleonic Code as the law of the land.

The question of legal sources was broached when a well-thought-of professor of civil law at the University of Heidelberg, Anton Friedrich Justus Thibaut (1772–1840), chose to dedicate an entire pamphlet to it. This pamphlet received a polemical response in kind from another eminent professor of Roman law at the newly founded University of Berlin, Friedrich Carl von Savigny (1779–1861). So began the dialectical opposition that characterized the history of German law throughout the nineteenth century. I should therefore sketch out for the reader the outlines of a controversy whose ramifications stretched well beyond the domain of theory into that of experience.

It is 1814. Thibaut releases an essay entitled *On Germany's Need for a General System of Civil Law*,[29] in which he outlines a very simple argument: he proposes that Germany should adopt a code, a tool in whose validity he sincerely believes, in order to provide a clear, certain and stable system of laws, as well as a useful spur towards unification. This would not mean the deployment of a French form of codification, still less the extension of the *Code Napoléon*, but rather a German code, drawn up by the best German jurists and incorporating the most praiseworthy elements of the common Germanic legal tradition.

Thibaut's work clearly reveals an underlying Enlightenment outlook – as can be seen by his extreme diffidence towards Roman law and its sedimentary practices, his enthusiasm for a general, abstract and rigidly unifying system of law, and his conviction that the better part of civil law is 'a sort of pure legal mathematics, over which no local factors may exert a decisive influence, such as the laws of property, the law of inheritance, the law of *hypothèques*, of contract and everything belonging to the general parts of legal knowledge'.[30]

That same year, Savigny raises an opposing voice in his seminal pamphlet entitled *On the Need for Legislation and Jurisprudence in Our Time*,[31] a book which has always been referred to for convenience by the first noun in the German title: *Beruf*. But who is this Savigny? Although this book has had to be sparing with personal details, given its summary nature, it behoves us to pause awhile over the life story of this German jurist, since he is undoubtedly one of the few examples of a single individual who had a notable influence on the development of European law as a whole.

Savigny was a conscious member of the historicist school – indeed he was the movement's leader, as his pamphlet in response to Thibaut demonstrates. He had close ties (of blood) to the highest echelons of German Romanticism, he maintained a friendly scholarly relationship with the Brothers Grimm, and with the cleverest researchers and prospectors in the field of German antiquities (institutions, customs, fables, mythology, language). Religious, aristocratic, anti-revolutionary and anti-Napoleonic, Savigny nurtured an intense belief in the historicity of human objects, with an accompanying intense respect for the past and tradition. Savigny was also ferociously hostile to bourgeois individualism and to the artifices of natural law which helped legitimate that belief system; he always sought to assert the collective and communitarian values which fostered the customs and traditions that represented the legal system's spontaneous basis and primal form.

The origin of the law – Savigny writes in his 1814 treatise – is in the 'common beliefs of the people', much like language, since 'the law, like language, lives in the consciousness of the people'.[32] Because of this, Savigny could not but be opposed to any idea of fixing the law into the immobility of general and uniform commands, to the idea of casting them in a text unaffected by the influences of historical change. The law, as a form of living history, would suffer the most unnatural mutilation under such a system. Instead, as Savigny comprehensively demonstrates, the law's principal source is custom, which is an expression of the community and in which historical continuity may be maintained. The primacy of the written law in certain historical contexts is thus merely the sign of their decadence and complacent corruption.

Savigny concluded that the German nation should avoid codification at all costs. Instead, he saw it as natural that there should be a process of refinement of the law during the course of a civilization's history. The reordering of the rough and spontaneous facts of custom would naturally be given over to a class of experts, the jurists, who would marshal these facts using scholastic categories and technical inventiveness but who would nonetheless be nothing more than the people's qualified representatives – allowing the production of law to maintain its flexibility. In Savigny's view, Roman law itself should be seen in the same light as German customary law because of its longstanding reception amongst German jurists and the development of the *usus modernus*.

It is important to mention the debate between Thibaut and Savigny here because it worked to sum up the competing proposals and motivations in the arena of legal sources and made these available to the German public: on the one hand there were those advocating codification; on the other there those who respected the law's natural malleability that stemmed from its two sources, custom and legal scholarship, with the work of the scholars interpreting the complexities of custom and thus representing the will of the people.

Savigny won out: his suspicion of codification undermined efforts to have Germany adopt a code, particularly because he spoke the language of Germanic nationalism and tied his rejection of the code to his resistance to the expansionism of the hated Napoleon. Germany would not inaugurate a civil code until 1896; instead the dominant force guiding German law of the nineteenth century along its newly unified trajectory was that of legal scholarship, as we shall now see. This trajectory was to a large extent shaped by Savigny's ideas and those of his followers. There is thus more to be said about him and about his long and complex intellectual endeavour.

The first, historicist, phase of Savigny's writings – epitomized by the pamphlet of 1814 – is followed by a second, dramatically different, one. In 1840, Savigny started to draft a work that would occupy him for the next decade, a work which can be called, without hyperbole, one of the most influential works of legal writing of all time, since its qualities have made an impression on generations of jurists both within Germany and without, as its many translations into foreign languages demonstrate. The work's name is *The System of Modern Roman Law*, a wide-ranging study that, despite its breadth, seeks only to sketch out a general theory of civil law.[33] As I do throughout this book, I shall attempt to rehearse here its essential message and to place it, without reduction, within the trajectory of Savigny's thought.

Savigny's work sets out to provide a framework, or at least some kind of organization for the extreme fragmentation of the many customs, an endeavour that can succeed only within the context of a wider structure of thought that reasons from the particular to the general and the abstract, allowing for the reconciliation of conflicting details. It is striking that this intimately reasoned system is seen as the creative construct of the jurist, using a rigorous and pure body of concepts and terms. Savigny's system is an impressive means of producing pieces of legal reasoning, crystals of logic that survive the test of time. One can also perceive a re-evaluation of the state and of the written law, alongside the central role played by legal scholarship.

So are there two Savignys? I would say that there is rather one Savigny who remains very consciously engaged with the legal historical developments and new requirements during the quarter-century that separates the *Beruf* and the *System*, and responds accordingly. German society had become more complex, the old ruling classes had completely declined and the entrepreneurial bourgeoisie that would head the expansion of the economy after the middle of the century had begun its rise to prominence. Meanwhile, the Prussian state became increasingly strong, something that affected Savigny personally, since the Prussian monarchs showed their faith in the jurist by appointing him to prestigious political offices, culminating with his nomination, in 1842, as minister for legislation. These factors cannot but have an impact on the *System*, which is written 'for the time in which we live' – it is not Roman law, but rather modern Roman law.

The ideas of the *Beruf* are not repudiated, therefore, so much as decisively updated. The two reflective moments of 1814 and 1840 share the fundamental choice of Roman law as the only model capable of giving them coherence. Roman law serves as an authentic example of 'modern customary law' – the supreme guarantor of the connection between life and scholarship: because it has penetrated the consciousness of the German people, it can therefore help to translate the network of customs into a network of academically rigorous laws. In Savigny's view, what is needed in 1840, much more than the granularity of facts, is a thought-out vision of how to organize those facts: something that cannot be invented artificially but instead must be discovered and described as an articulated series of connections that is laid over the historical reality. Reason is thus construed as a dimension internal to history itself, as classical Roman law itself demonstrates, since it remains the reading par excellence of the world of things.

Indeed, Savigny insists on the historicity of his system, which has a pyramidal structure, but one that does not ignore the base that supports the apex.

The multitude of *legal relationships* gives rise to the prime unifying entity: the *legal institute*, a composite of relationships that can stand for 'property', 'sale', 'intestate succession', etc. This complex of legal institutes gives rise to the *system*. The legal institute is thus the umbilical cord that connects the system to history and yet is itself a reduction of lived experiences to an underlying unity. Savigny is here an exponent of Roman law and of legal scholarship, who limits himself to placing a stronger emphasis on arguments that were already present in the *Beruf*; nonetheless, he cannot help underlining the need for general and abstract ordering systems to govern a complex society.

Savigny is not so insistent on the continuity in his vision of the role of the state and of the written law, and his clear re-evaluation of this area and his shifting of the load of the social, political and legal order towards state-based institutions is in manifest tension with his earlier ideas. The state is now 'the highest degree of legal creation', and Savigny argues that 'the law is the means by which the people's rights are asserted' and that the law-maker should be seen as the 'true representative of the spirit of the people'. Were it not for the phrase 'spirit of the people', this last quotation could almost be taken from the writings of Rousseau, a figure whom Savigny cordially detested.

From the point of view of this book, we are not so much interested in the detail of the relationships between the differing phases of Savigny's writings as in teasing out the wider implications that Savigny's work and the tensions within it have for the future of the law in the German-speaking lands. The historicism of the *Beruf* becomes the standard-bearer for the anti-codification movement, with the consequence that the German Civil Code does not come about until many years later than in comparable countries. (The *Beruf* certainly cannot be given all the credit for this fact – it is rather one force amongst many, such as the Germanists who privileged the study of indigenous customs.) The direction the German legal system took thereafter, and the imaginative legal machinery that came with it, allowed a distinctive form of Germanic legal scholarship to grow up in the legislative void and to take on its own specific shape. That shape will be the topic of the next portion of our discussion.

There can be no doubt that Savigny's later works enact a change in values: in the perennial dialectic between system and history that always afflicts the law (not to mention jurists), Savigny tips the scales notably towards the system. This comes about by his demonstration of the constructive capabilities of legal scholarship when it is freed from the demands of a legislative structure. As such, Savigny's work functions as a great storehouse of concepts, categories and rational ordering schemata.

Savigny's later works can therefore be seen as the inspiration and the guiding figure of an influential current of thought that dominated Germanic-speaking areas in the second half of the nineteenth century – the current usually called *pandectism*. Pandectism is so called because of the tradition that had existed in Germany from the sixteenth century on of updating Justinian's *Corpus iuris*, or *Pandecta*. The term implies a community of jurists – especially civil lawyers – who dedicated themselves with imagination and intelligence to constructing an abstract system of law, in accordance with the wishes of the middle classes who were now prevalent in Germany as well as elsewhere. This abstract system was organized into logical categories which were specifically designed to be free from contamination by the contingent economic or social facts. The abstract nature of the law created by pandectism easily gave rise to a system characterized by an absolute logico-deductive rigour. This was not so much the dominance of the system over history as a system entirely free from history.

Despite his formal protestations of fidelity to the tenets of the historicist school, the later Savigny most certainly provided a model for the exponents of pandectism, but an even more potent model was offered by one of Savigny's pupils, Georg Friedrich Puchta (1798–1846), a man imbued with an extraordinary aptitude for logical reasoning. In Puchta's hands the law became a sort of architecture, with Justinian law taking on the role of its unparalleled and ahistorical foundation. Puchta and his followers talk constantly of 'juridical constructivism', of the purity of principles and ordering schemata, and even of 'dogmas', that is to say a collection of unchanging truths that do not suffer time's decay.

These jurists, of whom the most noted was Bernhard Windscheid, were sworn enemies of natural law but did not notice (or wilfully ignored) the fact that they were producing a nineteenth-century version of the natural lawyers' work, aping their methodical approach and reaching the same essential conclusions: the technique of reasoning from models; the abstract descriptions of meta-historical subjects and relationships; the understanding of the law as private law first and foremost; and, most of all, the underlying ideology of individualism. Pandectism thus meant the identification of the law with formal order – indeed extreme formalism was its defining characteristic.

Finally, I should explain the reference above to the understanding between pandectism and the bourgeois order, and give such details as will make my meaning clear to the reader. The example of the juridical construction of private property will serve us well. The first codes that we studied showed some traces of the previous composite model of private property: rights of property understood as a series of swathing powers enveloping a thing and, like all things wrapped in swathing bands, able to be unwrapped or wrapped further. With pandectism, we arrive at a complete and coherent modern understanding of property: an all-embracing right rather than the sum of different rights. Property now reflects the indivisible individual subject with his liberty, and so cannot be modified by limitations to real property rights (such as easements requiring passage or access to water), but only compromised from without by the forces of an external reality from which it is completely estranged. (For comparison, take the title of the second book of the *Code Napoléon* that we examined on pp. 89–90 above.)

The Germanic Foundations of the Discipline of Public Law

As the reader will recall, Savigny's polemic of 1814 against Thibaut and against the idea of the immediate need for codification in Germany was entitled *On the Need for Legislation and Jurisprudence in Our Time*. The essential word here that has come to define it in the subsequent legal tradition is *Beruf*, which I have translated as 'need' but which could also mean 'calling'. The new century is thus seen as having a 'calling' for legislation and jurisprudence; in effect, Savigny is arguing that the new century will be a 'century of law': one that will seek refuge in the law, one that will place its trust in the law, one that will seek its clearest expression through the law.

Savigny's clarion call for jurists not to hold back from setting up new legal constructions was motivated both by pride and by understanding. The area to be covered by the new system was civil society and the fabric of everyday life – private law. In 1814 the law was still primarily a private affair: the 'public' was an extra-legal reality that was studied by political scientists and sociologists, whilst jurists concerned themselves only with excessive invasions of political power.

The jurists of the nineteenth century followed Savigny's call, offering an edifying example of a legal community committed to the construction of a system of private law that could adequately discipline German society's total economic transformation. The Savigny of the *System*, in the 1840s, has not rethought this model but there are, as we have seen, some innovations, demonstrating that the passage of time has not been in vain. Alongside customary laws and legal scholarship, the law takes on a greater role, and the state receives new attention.

The fragments of message lodged in amongst the wrinkles of the *System* will in fact lead to a proper cultural and technical project pursued by a character who could not have been left out of this book: Carl Friedrich von Gerber (1823–91). Gerber was a singular figure, not least because of the complex trajectory he travelled in order to become a jurist: at an early stage of his scholarly career he gave abundant proof of his capabilities with a systematic treatment of German private law, but he also had the virtue (and the burden) of possessing an extraordinary historical sensitivity and awareness. The middle of the nineteenth century was a turbulent period in German political and social life: the liberal adventure of 1848 was conceived and then repressed, and Prussia continued its unstoppable march towards both a more rigid state structure and an ever larger share of German-speaking territory.

Gerber considered that his era possessed the 'attributes of a transitional period', as he wrote in a revised edition of his most famous work, the *Features of Public Law*. Gerber's consideration is not a detached one, indeed he is an eager participant in this transition. Some have accused Gerber of opportunism, of legitimating German princes' reactionary measures that abolished sworn constitutional freedoms. But I am interested here in only one significant detail: that a private lawyer like Gerber, who absolutely shared Savigny's analysis of society and adhered to his call to legal constructivism, could argue that the moment had arrived to put aside his disdainful attitude of complacency towards the public sphere and advocate the incorporation of this sphere within legal scholarship's constructive enterprise. The feeling of a need for change was increased by the 'febrile impatience' that coursed through German cities in 1848 and shook Gerber out of his mournful diffidence. This change would not be sparked by the spontaneities of pamphlets, as had happened in neighbouring France, but rather through the slow, insightful and enduring structures set up by jurists, which imbued the new principles and rules with the wisdom of customary law in a manner similar to the English common law, much beloved of Gerber.[34]

On 6 April 1851 one of Germany's great jurists, a man who would later be famous throughout Europe thanks to his works written for a general audience, heard of Gerber's proposals and was moved to put pen to paper and write an extremely eloquent letter of which I shall transcribe a fragment here. The letter writer was Rudolf von Jhering (1818–92), a Roman lawyer and later the co-founder, along with Gerber, of a learned journal devoted to the 'dogma of modern-day private Roman and German law'.[35] Jhering writes, 'You speak of constitutional law? But my good friend, I would do no such thing; it is not yet the time to confront constitutional law. […] Keep to your private law, write your doctrinal study once and for all. […] You have an important interest in the dogma of private law'.[36]

But Gerber was acutely aware of the needs of his own time, and paid Jhering no attention, embarking on a pioneering work, which not only gave a legal meaning to the idea of the *public sphere*, but used formal legal concepts taken from the teachings of private law to set up a similar system for public law.

There were two stages to this constructive progress: a book-length study of the problem, *On Public Laws*, in 1852 and, in 1865, the *Outline of a System of German Constitutional Law*. The former work constitutes a scholarly experiment providing a preliminary sketch of the issues; the latter is a plan for a 'system'. The *Outline*, known by its German title the *Grundzüge*, is a mature and forensic consideration of the need for a constitutional system, but it also functions as a positive appraisal of the unitary Prussian state as it then stood – not yet the Reich of 1871 but already a fully fledged realization of a powerful and unified state apparatus.

Gerber faithfully chronicles the people, the *Volk*. In the wake of the suggestive contributions of Savigny and the historicist school, the people had long maintained their status as the original basis of society, autonomous under their historical carapace of values, beliefs and customs. Now they were totally absorbed into the unitary structure of the state, helping to imbue it with ethical value and thereby acquiring a higher ethical status for themselves. Indeed, the *Grundzüge* begins by stating that 'the state gives legal discipline to the collective life of the people. The state allows the people to be recognized and to have legal worth as a single entity'.[37]

Gerber goes on: 'if we consider the state from a legal point of view, we see straightaway that, through the state, the people may grow together into the consciousness and the discipline of will that a society of law requires'.[38] He concludes, 'the personality of the state can serve as public law's point of departure, its nucleus; the personality of the state underpins the possibility of having an objective system based around a single worldview, whilst, at the same time, it provides its ordering principle'.[39]

The most significant achievement of Gerber's dogma is the transfer of the notion of personhood from the individual to the state – an idea which must have seemed intuitive to a private lawyer like him. The results of this transfer were twofold and equally significant: first, the establishment of the perfect unity of the state, which is therefore reduced to the will of power; second, the complete separation between the state, which is a purely collective entity in which individuals are subsumed into a greater unity, and society, which is the realm of individuals' activities and liberties (amongst which the first is, of course, the freedom to own property). And so began the double function of the state: interpreter and underwriter of the common good – the 'supreme guardian of the collective interest', and the guarantor of the public rights of individual subjects, although these rights were to be seen as derivative.

Gerber's state is undoubtedly a state of laws, in that it is defined by the production of law, and the law defines its limits, but this law is nothing more than legislation, or to put it another way, it is nothing more than the manifestation of the state's will, no matter how impersonal, general or abstract that will might be. Thus we can see that the Gerberian state is responsible for setting its own limits – and so the state of laws reveals its true authoritarian colours.

This authoritarian character applies throughout the most important German tradition of public law scholarship. Gerber still finds some place for jurists, who are called upon to think up suitable legal structures and to construct legislation, but by the time of Paul Laband's (1838–1918) *Constitutional Law of the German Empire*, published in 1876, legal scholarship is so subsumed within the apparatus of power that it has been reduced to the lowest rank, that of creating orthopaedic braces for the state's wishes. The job of a scholar of public law is now to describe the existing situation and to find a means of defending it at all costs.

The field of public law as a whole is somewhat more fragmented than this account might suggest, although the dominant outlook, in my view, is still that which passes from Gerber to Laband. For the purposes of this book it is enough to note that, alongside the construction of a system of private law, there was a parallel effort in the field of public law using a formal and abstract system of reasoning. The result of this was a formalistic system of public law; it is symptomatic that the originator of this system was the private lawyer Gerber. Just as the abstract nature of private law safeguarded the German bourgeoisie's bloated interests, becoming a formidable protective carapace, so the abstractions of public law safeguarded the new legal personality of the state.

Pandectism and Germanistics: Towards the German Empire's *Bürgerliches Gesetzbuch*. A 'Scientific' Code: The BGB of 1896–1900

In 1881, with Germany now politically and economically stable, the journey towards a civil code began. A commission was set up, and by 1887 it had proposed a preliminary version of the new code.

Although both legal scholars and practitioners participated in the commission, the final product was characterized by abstraction and by a decided tendency towards individualism, a complexity of language and doctrinaire provisions. It was thus difficult for the average citizen to access, although the average citizen was not, in any case, the addressee the writers of the code had in mind – they were writing rather for the legal expert in his robes. The commission was headed by the charismatic figure of Bernhard Windscheid (1817–92), the prince of German pandectists, who made sure to imbue the paragraphs of the code with the worldview of pandectism. He met little opposition in this effort, since the preparatory works had already demonstrated the complete adherence of all the members of the commission, even the practitioners, to pandectism – their education (some might say indoctrination) in the universities of Germany had initiated them fully into the dogmas of the new cult of legal construction.

The effect of the draft code, however, was like that of a rock thrown into a dovecote: it set off a volley of polemical ripostes. I shall mention only the two most significant ones here, the first arguing from a cultural point of view and the second taking an ideological stance. Otto von Gierke (1841–1921) was a Germanist who had been working with increasing renown on indigenous customary law – a more marginal and discrete field with respect to the official status of the Roman lawyers. Von Gierke invokes this resurgent social and communitarian national tradition, which had been deliberately sidelined by bourgeois individualism, to make a head-on attack on the draft code which scathingly contrasts the artificial and selfishly bourgeois abstractions of pandectism with the solidity of the *Deutsches Recht*.[40] Anton Menger (1841–1906), meanwhile, who was a professor at Vienna in civil procedural law and a believer in solidarity, although not in the class-based politics of Marx, attacks the document for its indifference to the needs of the non-property-owning classes.[41]

In 1890 a second commission was set up, and in 1895 a new draft was produced, followed by a third the year after. In 1896 the code was finally promulgated, and it came into force in 1900.

The *Bürgerliches Gesetzbuch* ('Civil Code'), generally known by the abbreviation BGB, is certainly a code in the legal historical sense of the word described above. It asserts its exclusive jurisdiction, reaffirms the validity of positive law and charges judges with observance of state-made law. The BGB's interlocutor remains the judge; its style is thus highly theoretical, rooted in the concepts, techniques and rigorous purism of the most refined pandectism.

This, then, was the hidden protagonist of the code, and so it was Windscheid who triumphed at the end of nearly two decades of sometimes bitter debate. The cult of the general and the abstract dominates, and any hint of particularism or case law is scrupulously avoided. The legal institutes are fixed in a conceptual framework of extreme logical rigour. The pandectist cult of abstraction reaches its zenith in the so-called 'General Part'. This bears little resemblance to the scant few articles of the Napoleonic Code or the Austrian Code that deal with legal sources in general; it is rather an entire preliminary book, by way of preface to the code, that extends over fully 240 paragraphs. And yet it is nonetheless integral to the code: it is a true general theory of civil law, including large tracts of an even wider general theory of law, each punctuated by accounts of the essential conceptual unity between various legal relationships – a concern that is also treated at the beginning of the second book in paragraphs 241–432. A significant example, taken from the centre of the first book, is general theory of *legal negotiation* (*Rechtsgeschäft*), significant because such negotiations are the way in which the individual subject asserts his interests when entering into relationships with others. Legal negotiation is thus the most jealously guarded terrain in an individualistic construction of civil law.

One typical feature of this codification is the recourse to the 'general clause' as an expedient which serves to free the code's abstract rules from becoming entangled in the minutiae of concrete cases. 'General clause' means a directive made by the legislator to the judge that authorizes him to have recourse to notions and experiences beyond the boundaries of positive law. Thanks to the references to generalities such as 'good faith', 'good custom', the 'habits of commerce' and the 'diligence of a good father of a family' (some of the more common general clauses), the judge takes on more responsibility. These provisions included by the code's drafters, and circumscribed within tight limits, open a release valve in the formal structure of the codified system that guarantees a greater degree of adaptability in the general rules without relenting on the principle of exclusivity.

One final point to note: the BGB is clear in its diffidence as a bourgeois legal document towards the social and the collective; it repeatedly affirms its rigidly individualistic outlook. At the economic level too, the BGB clearly sets up the individual entrepreneur as its point of reference.

A Peculiar Codification: The Swiss Civil Code of 1907

Over the course of the nineteenth century, several Swiss cantons had engaged in very distinctive programmes of codification. Pre-eminent amongst these was that of the canton of Zurich, whose *Privatrechtliches Gesetzbuch* ('Private Law Civil Code' 1853–5) was the work of one Johann Kaspar Bluntschli. Nonetheless, the Swiss Confederation would have to wait until 1907 for a unified system of private law, even though a unified federal constitution had been in place since 1848.

In Switzerland too the birth of a code was a protracted affair with several stages taking more than twenty years in total – although there was never the bitter controversy that bedevilled the inauguration of the BGB in Germany. In Switzerland the process was simplified by two factors: the codifiers were working against the backdrop of a political sphere in which direct democracy already played an established governmental role; and in reality there was only one codifier, a faithful exponent of the Swiss juridical tradition who was able to give the code its own specific character. There is no other code that is so clearly written by a single identifiable 'author'.

That author was Eugen Huber (1849–1922), a professor at the University of Berne, and a jurist of extraordinary learning with a truly European perspective because of his dense network of contacts. He was, however, not a Roman lawyer but a Germanist with a deep knowledge of Swiss legal history. His proposals were favourably received because they were rooted in the solid earth of national tradition. Not for Huber the BGB's doctrinaire posturing, refined but impenetrable language and conceptual abstractions devoid of practical applicability.

Because of its singular character, the Swiss Code deserves special mention here. Its defining characteristics were as follows: the interlocutor is the individual citizen, meaning that the document's language is plain, concise and intelligible to all and that it makes consistent reference to daily life – the frequent use of proverbial expressions is a good example of this; there is no general section; the entire code is founded upon a sense of solidarity, which leads to a much more concrete approach to the law in which inequalities of fact are acknowledged and positions of social or economic strength and weakness are considered legitimate subjects for the legal order to cover; it gives significant space to collective structures, a reflection of a vibrant socio-economic reality in the Swiss plains and mountains; commonly accepted concepts such as 'common sense' and 'good faith' are given significant space in order to infuse the necessarily closed system of the code with the oxygen of everyday habits and practices.

But the most important feature of the Swiss Code is the fact that it makes the judge the protagonist of legal life. This is a courageous choice on Huber's part and it very much sets the Swiss Code apart from its contemporaries. In opposition to the Enlightenment-era suspicion of the particular, the judge is placed in the vanguard precisely because he is immersed in the details of daily life, in the concrete facts of history in which the law is truly inscribed. Here are the second and third sub-sections of Article 1, translated from the official Italian version of the code: 'In cases unforeseen by the law the judge shall decide according to accepted custom; should this also be absent he shall act as legislator and create his own law. In this he shall follow the best teachings and jurisprudence available.' The old Enlightenment prejudice against judges seems not to have reached the valleys and mountains of Switzerland.

Over-legislation in France. The 'Praetorian' Power of Practical Jurisprudence. Intolerance and Proposals for a New Code: Raymond Saleilles and François Gény

In 1904 the centenary of the Napoleonic Code was celebrated with great pomp in Paris. There was, of course, the inevitable wave of complacently nationalistic eulogies, but there were also several haughty lectures given by civil lawyers and judges expressing the somewhat strange opinion that the greatness of the code consisted in the generic nature of its prescriptions. It was of course that very generality that had necessitated the supplementary prescriptions made by academics and judges.

The passage since 1804 of the better part of the nineteenth century, a period full of new challenges, had demonstrated the insufficiency of the code as a piece of legislation and the urgent need for reform. The gaps were not only in the economic ideas of the codifiers (as Pellegrino Rossi had pointed out in the 1830s); the code was simply not fit to oversee the legal order of a country undergoing extremely rapid growth. A retrospective account of all these flaws would be given in the early twentieth century by a French jurist-practitioner, Cruet. Cruet's book, entitled *The Life of the Law and the Impotence of Laws* (1908), efficiently summed up the issues that had divided nineteenth-century France: the tension between the codified text and lived experience and the 'impotence' of that text in the face of an ever more complex historical reality.[42]

The cumbersome nature of the code as a legislative tool meant that its rule became ever less bearable as time passed, and so, in the last decade of the century, we see a shaking off of this passive and deferential attitude towards the code and the beginning of a new process of legal construction. It is interesting to note that the movement aligns itself ideologically with the interpretative labours the judiciary had been carrying out – an extensive and undoubtedly innovative programme that was nonetheless slow and discreet, judges' work took place between the cracks of the code, but went beyond the code and sometimes even against it. These innovative civil lawyers admired the 'praetorian style' of the French judges and especially of the Conseil d'État. ('Praetorian' refers to the ancient Roman *praetor*, the magistrate in charge of adapting the legal system to new challenges.[43])

Following in the footsteps of unheralded but effective verdicts made by judges, case by case, day by day, over several decades, there emerged in the 1890s a cohesive and battle-ready group of young jurists. They had almost all recently written doctoral theses and were thus committed to research and to climbing the hierarchical ladder of the French university system.

Let us touch on some of the more burning problems that these jurists confronted with liberality, courage and cultural and social sensitivity – all of them problems that the code left unresolved, ignored or dealt with in an anachronistic and therefore unsuitable manner. I shall attempt to make clear what was bubbling on the stove, concentrating on the problems of the greatest ethical, social and economic import. These included: the abuse of rights, that is to say the attempt to constrain the liberty of the individual in the name of the community; the need for a means to assign debts, and thus to free French law from the ancient Roman conception of the debit–credit relationship as existing indissolubly between two specific individuals; the need for greater regulation in the area of labour and, more widely, that of civil liability – again an attempt to extricate French law from the old Roman doctrine of individual culpability for an offence and to set up a principle of objective liability instead.

It would be interesting to dwell further on this intensely reflexive moment in French legal thought, a period with many connections to the more general theory of law, but there is no space in this book. We shall limit ourselves to mentioning briefly two great figures who had a profound cultural influence on the future development of French law.

The initiator of the innovatory movement and the undisputed mentor of the young jurists was a professor of civil law called Raymond Saleilles (1855–1912). Even though he was an active participant in the making of new models of law, Saleilles had the virtue of having experienced at first hand the problem of the sources of law in France at the end of the century, with the venerated code bobbing aimlessly on the ebb and flow of contemporary legal studies.

Saleilles began by denouncing the outdated syllabus of the faculties of jurisprudence in French universities, and their slavishly exegetical methods that left the students with no awareness of the realities of the civil law system. Both the syllabus and the methods were strategically adapted to bolster the dominant legal mythology and especially the cult of the codified text as sacred scripture. In Saleilles's view, it was this crystallization of the law into an authoritative text, and the consequent blinkering of young civil lawyers so that they studied nothing but texts, that were responsible for the stagnant climate of French law at the time. Saleilles thought French law needed to create an opening in the suffocating wall of codification and put itself in contact with daily life. The exercises in legal logic based on authoritative texts were all very well, but they needed the counterpoint of an immersion of such texts in the lessons of history, economics and politics. There had to be a reckoning with social change.

Saleilles was a fervent Catholic but nonetheless a devoted modernist, that is to say an adherent to the theological position that Holy Scripture can be judged according to historical change and its prescriptions modified according to historical circumstances; he extended this attitude to the immutable pseudo-sacred text of the civil code. His vision of the legal interpreter grants such figures considerable freedom: the authoritative legislative texts remain the centre of attention but the interpreter can look 'above and beyond the texts', using a 'method of historical adaptation' and setting up a 'more flexible and elastic system'. These changes will lead to the result desired by Saleilles: *l'assouplissement du texte*, the 'softening' of a text that until that point had been believed – absurdly – to be rigid and immobile. For Saleilles, the Napoleonic Code had been drafted in such a way that it would be functional only with the help of later generations of sensitive jurists capable of adapting its generic prescriptions to the changed requirements of their own time.

Of course, Saleilles is not so foolish as to ignore the surrounding legal climate: he does not advocate a complete break with the past – as does the free-law doctrine considered in the next section – such a point of view would have been suicidal in *fin-de-siècle* France, where not only the law but now the code too was idolized. Moreover Saleilles's proposals contain some slipperiness: he stipulates the judge may stretch the text but only within the 'frame' that it offers. Such a statement cannot help but provoke in the reader the justifiable question: what happens if the new facts do not fit the 'frame' and cause it to shatter?[44]

But we are not interested in the inconsistencies in Saleilles's ideas; instead we are examining him because it is instructive to see the unease that the continued existence of such legal obscurantism provokes in this educated and open-minded scholar. Also of interest to us is his research into new means of giving value to other sources of law – such as scholarship and practical legal experience – that were more capable than the legislator of giving voice to that complex reality that is the law, not to mention his insistence that law is life, that it is a very changeable experience and that it is crushed, not protected, if legislation is allowed to monopolize. Finally, as well as his sincere desire to open up wider horizons and to broaden the minds of late nineteenth-century French jurists, we should also point out Saleilles's prescient acceptance of comparative law; as we shall soon see, he was responsible for planning and organizing the first international conference on comparative law in 1900.

Another civil lawyer who was equally uneasy with the legal status quo is François Gény (1861–1959), a well-respected scholar of both the nineteenth and the twentieth centuries.[45] Just like Saleilles, Gény was courageous enough to confront the two questions he saw as urgent: the underlying problem of the sources of law, as well as the narrower issue of interpretation. Unlike

Saleilles, however, who preferred to disseminate his work in numerous brief but rich contributions, Gény wrote a powerful work full of critical analysis and suggestions for systematic reform, entitled *Methods of Interpretation and Sources in Private Positive Law*, which was published in 1899, at the very end of the century of codification.[46]

The primary motivation for this work stemmed from the 'infinite complexity and the incessant mobility of social life', which cannot but lead to a rejection of the 'fetishism of the written and codified law', since 'the written law is nothing more than the imperfect revelation' of the infinite riches of the law, and is thus an insufficient source to govern the legal universe alone.[47]

Gény's solution does not draw on Saleilles's idea of stretching the authoritative texts, since he views this as potentially damaging and in any case impractical. Instead, Gény tends to confine the relevance of legislative texts as much as possible, taking two premises as his starting points. The first is that a law is a command, the expression of a will: it is therefore not possible to 'stretch' a law, but at the same time it must be specific in what it prescribes; the subject of a law must thus be redefined as closely as possible to that which was originally intended, without any possibility for extension. The second is that the code is not a complete system with a few gaps, but, on the contrary, a necessarily incomplete system. From these two premises, it follows that the interpreter is limited only insofar as the express and specific word of the code binds him, whilst he remains completely at liberty in any part of the broad areas in which legislation either has not intervened or has intervened only in a generic or a confused manner.

The Free Law Movement: A Legal Form of Modernism

Saleilles and Gény were only the most visible exponents of a more widespread European discontent with codification. In France this feeling may have been limited to a group of impatient and brilliant youngsters, but it was much more prevalent in German-speaking areas, where, from the 1870s onwards, even the more grizzled legal scholars were coming round to this view. There were a number of motivating factors for this shift: Austria's Code was already old and outdated, whilst Germany was beset by draft codes that were characterized by total abstraction, although the extreme conceptualization of pandectism was now the dominant ideology in Vienna too. Given these considerations, it is instructive to look at the maturation of the thought of the great German jurist Rudolf von Jhering, who moves from a pure and abstract vision of the law to place a much greater emphasis on the external forces and interests that shape and guide the law.

The discontent felt in France and Germany would seem to have a common root: the excessive separation between the law made by legislators and scholars from the law as it is lived. The malcontents also shared a common goal: that of finding some way to historicize the law effectively.

In the first decade of the twentieth century the accumulation of unhappiness overflowed, and a movement known as the free-law movement was founded (the term is a related to two German terms: *Freirechtsbewegung*, 'free law movement' and *Freirechtslehre*, 'free-law doctrine'). The people who made up the movement were relatively varied, but their inspiration was one. Free law was more of a movement and a doctrine than a school, that is to say scholarly jurists were joined by practitioners – judges and advocates alike – all united by the sincere and passionate desire for legal reform.

The best expression of the beliefs of the free-law movement can be found in the following texts: various writings by the intelligent and well-educated lawyer, Ernst Fuchs (1859–1922); the essay *The Free Discovery and the Free Study of the Law* written in 1903 by Eugen Ehrlich (1862–1922), a professor in the far eastern parts of the Austro-Hungarian empire and the future founder of a whole new discipline – the sociology of law; and the well-known pamphlet *The Struggle for Legal Theory* from 1906, written by the legal historian Herman Kantorowicz (1877–1940).[48]

The essential feature of the movement is easily grasped if one pays attention to the heading under which these singers of a new song are gathered in France, Germany and Italy: *modernisme juridique*, *juristischer Modernismus*, *modernismo giuridico* ('legal modernism'). The movement thus is seen as putting forward a new outlook on issues such as the role of the wrriten law, the relationship between statute law and the law as a whole, the role of interpretation, the application of the law and, therefore, the role of the judge. This new outlook contains many developments of ideas already present in Saleilles and Gény. It was certainly not an anarchist movement as many of the *bien pensant* members of the legalistic majority held, disturbed and irritated as they were by the starkly polemical pages of the free-law movement's supporters. The nucleus of the free-law movement's reappraisal of the law was made up of its indictments of three aspects of the contemporary legal sphere: the overemphasis prevalent at the time on statute law as the sole source of law; the undiscriminating cult of the law; and the belief that written law alone could express all the riches and complexity of the legal sphere.

It was this legal idolatry that provoked the anger of the free-law movement's supporters, who were convinced that the law had been in some way confined to a crushing orthopaedic support, unnaturally immobilized when by nature it was mobile, and so unwisely sacrificed on the altar of a drastically reductive operation of codification. To their eyes it was risible that all problems could be solved with the magic wand of the code, and particularly with the half-truth of its completeness – which they saw as nothing more than a glaring falsehood intended to guard the state's monopoly on the production of laws.

Kantorowicz's words on Article 4 of the Napoleonic Code are illustrative: he sees the text for what it effectively desires to be – a bolt closing the cage of the code – and exposes as false the idea that a code can be an organic and complete system whose few lacunae, where these are admitted, may be filled by recourse to other points within the code. For Kantorowicz, the article puts forward such a proposition in the service of the legalistic culture. Yet the truth is exactly the opposite, and Kantorowicz is unflinching in his riposte: 'we can now say honestly and with some confidence that the gaps in the written law are no less important than the words!' The code, in effect, is as much a gap as it is a system, not only because there are lacunae, but also because of its imprecisions, generalities and unsatisfactory approximations.[49]

The written law was thus insufficient, and it could not be otherwise; it therefore could not be left alone and unchanged. This did not mean it needed to be torn down, but merely that it had to be accompanied by the *free* interpretation of scholars and the *free* application of judges. For Kantorowicz, the role of the judge especially required re-evaluation: the judge was seen as the guarantor of the law's historicity, that is to say of its capacity to keep pace with the dynamism of social change – the model here was the English judge, who is the prime mover of the ever-changing common law. It is the role of the judge in particular that distinguishes the free-law doctrine from the mechanistic solutions of the Enlightenment that require the judiciary only as cogs in a logical and deductive system of gears. On the contrary, in the new vision, the judicial decision is imbued with the qualities of free will, intuition, common sense and justice, whilst his application of principles of equity is no longer the nightmarish proposition it was for the legal idolaters.

At the heart of the free-law doctrine is the recovery of the evaluative aspect of interpreting and applying the law, as well as the reintegration of the realm of facts and history into the law. The primacy of the individual and of the particular is once again asserted, and these become the mirror that gives concrete form to the abstract and general propositions that, by themselves, can create only artifice.

Legal Solidarity of the Late Nineteenth Century: The So-Called Social Laws

The second half of the nineteenth century sees the triumph across a large expanse of central and western Europe of mature capitalism, particularly in Britain, Germany and France, all of whom still possessed an intact colonial empire. Capitalism brings with it its most typical concrete manifestation: the large enterprise. Such organizations seek to maximize output and profits by making use of technology and labour, but they were also responsible at this time for bringing about two historical realities with considerable socio-legal ramifications. The first is the primacy of the machine over the individual labourer; the second is the necessary concentration of a mass of workers in one place over a sustained period of time.

The first of these circumstances makes clear on a macrocosmic scale a fact that, in the context of a small enterprise using only limited technological apparatus, might have remained both less true and less obvious. Work-related injuries, the fragility of the human person in the face of industrial machines and the issue of safety in the workplace all contributed to forming an impression of the supremacy of the machine over the person.

The second circumstance gives the multitude of exploited workers the chance to communicate with each other, to share their problems and their grievances, to begin planning common struggles. The large enterprise can thus be seen as the location in which the working-class movement first takes shape. This movement represents one of the most significant developments of the last decades of the nineteenth century: with the continual stream of protests and demands from one side met by brutal retaliations from the other.

Outside of and indeed ferociously against the bourgeois sphere, the authentically socialist movements were making their advances. However, at this point in history they have only scant relevance to our project, since their effect on the law was minimal: they set themselves up as prosecuting a subversive agenda based on the class struggle and aimed at tearing down the entire bourgeois order as it had been constituted since the revolution of 1789. The protests that took place, meanwhile, can easily be seen as part of the drama of the bourgeois order, and indeed these did bring about some notable innovations. In fact the protests should be seen as advocating not *socialism* but *solidarity*.

To explain: a great part of bourgeois officialdom adopted an attitude of deliberate closed-mindedness with regard to the workers' demands, instead choosing to confront the protesters head-on. But there were also those who accepted, within strict limits, the need for dialogue with the expectations and demands of the proletariat. Such moderates wished to reduce some of the more excessive social inequalities of the era by allowing the state to intervene on an economic and social level in order to mitigate situations of obvious hardship. Bourgeois individualism could accept only the merest traces of such charity towards the classes most in need of help: there would be no socialism, only solidarity.

These developments could not but have an effect at the level of legal sources. Until then the legal order could count on a catch-all source, the ur-law of the code, but this had been revealed to have been drafted and implemented in a misguided manner and thereafter defended as if it were a set of general and abstract principles. The code's view of the human subject was thus an isolated one that bore no relation to the nineteenth-century historical reality but rather belonged in a museum of models of human individuality. In effect, the code was law that was unsuited to adapting itself to the demands of a particular time and circumstance because it was expressly designed not to prostrate itself before the base facts of daily life. There were those who called to have done with the code, and others who began devising a private social code based on facts and corporeal beings, unlike the austere *Code civil*, and there were those who talked in terms of 'special laws', that is to say laws that did not dismantle the skeleton of the code but rather were integrated with it in order to make it more specifically relevant.

The idea of the special law was not a new one, although the device had seldom been used so as not to impinge upon the codes' supremacy. But now there was another role for such procedures to fulfil: filling in the gaps in the codes by modifying some of their propositions. The state was so wedded to one social class because the storied articles of the codes were laws that the political class could bend to its purposes. The parliamentary debates of the time are full of callousness towards the workers and contain little by way of open-mindedness, lines of communication or understanding. But the solidarity movement forged a path forwards nonetheless and, from the 1870s on, there was an efflorescence of special laws – not an enormous one but not inconsiderable either. These were known as 'social laws' because of their goals; they were laws that dealt with the most urgent social issues.

These issues would be taken up by a sensitive parliamentarian or member of the government, after which they would become draft bills, often with a long, tortuous and often abortive path into law. But they might also be part of the paternalistic projects of a state whose wider position was predominantly absolutist. The emblematic example here is Bismarckian Germany. The end of the nineteenth century there saw a whole series of social interventions, beginning in 1869 in the North German Confederation with the *Gewerbe Ordnung* – a law on industry and labour. The official interpretation of the government of its law establishing insurance against sickness for workers is illuminating: 'the state must take care of its needier members to a greater degree than it has up to now. This is not only a duty required by common humanity and by the Christian faith, but it is also a necessary pillar of a conservative policy, in order to foster among the less fortunate classes, who are both more numerous and less educated, the belief that the state is a beneficent and indispensable institution'.[50] 'Social laws' can thus be seen to be an integral part of enlightened absolutism, indeed they can be argued to be an excellent technique for fostering conservatism, since they do not represent the state as ceding to the will of protesters but rather as asserting its strength independently. The German imperial government hoped to use such provisions to win greater obedience from its subjects.

The places most favourable to such social laws were Britain, Germany, France, Austria-Hungary, Spain and Belgium. The areas covered tended to include workplace injuries; children's and women's working hours; health and safety in factories; compulsory insurance against invalidity and old age; social security funds; and arbitration and reconciliation services to solve disputes between employers and employees.

I said above that large enterprises proved to be very fertile ground for the nascent workers' movement. This spontaneous movement was immediately the object of fear and every type of repression, since it represented the worst nightmare of a bourgeoisie that had striven to eliminate all types of non-state association. The assertion of a population of equal citizens was a doubly effective expedient: it theoretically extended rights to the poor as well, and yet the state and the citizen were left without responsibilities towards each other, ensuring that the weak stayed socially weak and therefore innocuous.

From 1791 onwards, when the Le Chapelier law had erased all social institutions, there were only solitary individuals: a solitude that rewarded those who owned and left the destitute even more naked than before. It was the triumph of the *I* as individual over the collective *I*. And yet the integration of the individual into collective structures was the only road to salvation for someone of working-class origins: the collective was the only way in which such an individual could abandon the abstract principles of equality that did nothing to help his situation and make up for the deficiencies of the non-existent individuality granted by the new laws by finding a protective niche in a wider structure.

As we approach the end of the century, we can see an ever clearer conception of the collective *I* for the working classes. Since we cannot cover all the various protests here, let us concentrate on one, which took place in France, inevitably the nation in which the political choices of the revolution were held most dear and most often seen as insuperable instances of historical progress. On 21 March 1884, a law was passed granting legal status to professional associations, including trade unions: all workers could now unite freely without the need for authorization from their organization.

One final note: this was also the moment at which co-operative associations began to represent a force. There were therefore specific laws governing the new social institutions: co-operatives and mutual insurance societies.

Legal Solidarity of the Late Nineteenth Century: So-Called Legal Socialism

So maybe we can talk in terms of 'social laws', although it must be kept in mind that the states that made the laws still had only solidarity as their intention. And it should be made equally clear that, although legal historians will often write of the turn of the nineteenth century as a period of 'legal socialism', we are still dealing with mere variations in the bourgeois universe rather than true socialist solutions. The terms *socialism* and *socialist* are much overused, sometimes carelessly, sometimes with a specifically pejorative nuance, and this can create many confusions.

In the early 1870s, 1872 to be precise, an association for social politics (Verein für Sozialpolitik) was founded in Eisenach in Germany. This association served as a meeting point for a group of university economics professors who were exponents of what is known as *Kathedersozialismus*, or university socialism.[51] Their reformist agenda went no further than a demand for decisive state intervention in the economic field with the exclusive aim of reducing what they saw as the excessive imbalance between capital and labour.

In the legal sphere the same sort of thing was happening across Europe in various different languages: this is so-called legal socialism. But where? How? And on behalf of whom?

If I am not mistaken, the first writers to use this term – and with a derisory undertone – were Friedrich Engels and Karl Kautsky, the faithful evangelists of the new Marxist gospel. The occasion was an article in 1887 in the Communist Party's organ, *Die Neue Zeit*, about the Austrian expert in procedural law, Anton Menger, whom they call an exponent of *Juristensozialismus* – a term which indicates the false socialism put forward by certain jurists, who, as lawyers, would not be able to avoid distorting and falsifying the liberating message of the words of Marx and his followers.[52] But, since Engels and Kautsky's article, the term has now transferred to both Italian (*socialismo giuridico*) and French (*socialisme juridique*), where it has lost the pejorative nuance.[53]

Menger's essay, to which Engels and Kautsky are responding, has already been mentioned here, and it was written as a polemic against the first draft of the BGB – a document which was completely enslaved to the pandectist ideals of the purity and abstraction of the law and of the future code. The essay in question is a sample of Menger's more complex and articulated body of work, which can be taken as exemplary of the contributions made by the so-called legal socialists from across the European legal sphere over the course of the late nineteenth and early twentieth centuries.[54] Given that we have already established the basic character of the movement as a whole, let us quickly sketch out the essentials of Menger's proposals. I should make plain, however, that legal socialism was merely a movement united by a common desire for solidarity, not a 'school', as it was composed of people from very different cultural backgrounds.

The proposals come from within the bourgeois edifice and certainly do not envisage tearing it down. They are, rather, ideas for modification, which deliberately depart from the late nineteenth-century conception of the state and of the law in an attempt to preserve them. In other words, Menger departs from the premise that the edifice has many faults and flaws but that these can be rectified. If something must be eliminated, it is the general and abstract form of law, which, since it only helps the wealthy, inevitably becomes a class-based law – a law for the minority only. Private law thus has a need to be made more 'social', that is to say there is a need to orient it more along lines of social justice, by making social rights the heart of the law, preventing exploitation, and stopping profiteering at the working class's expense.

One can see why Marx and his socialist followers might consider these ideas mere *Vulgärsozialismus* ('pseudo-socialism'): firstly it is misguided to attempt to bring about change by concentrating on the distribution of wealth whilst leaving the means of production untouched; it is equally misguided to try to effect redistribution using the instrument of the law – a bourgeois institution and one that will remain so. An Italian socialist, Claudio Treves, a journalist and member of the lower house of the Italian Parliament, wrote lucidly of the issue in the party magazine, *Critica sociale*, in 1894. Treves says: 'We are awaiting the reform of private law that will result from this flourishing organizing movement amongst the workers, which is taking majestic shape out of the great international uprising of the proletariat. [...] The work of jurists can add almost nothing to this effort. Socialism can only come about through the actions of the interested parties: therefore, I repeat, the para-socialist movement that has convulsed the keepers of the law and that still from time to time makes itself heard abroad is completely sterile.'[55] It is hard to think of a stronger, franker, more dismissive way of putting it. Treves sees so-called legal socialism as a sentimental attitude drifting through the elite precincts of the jurists. It may have been thus, but for others legal socialism was something quite different: for some it was a clever strategy to prop up conservatism, for others a generous nod towards solidarity. Whatever it was, it was not socialism.

However, the movement had one merit that must be recognized: it helped Europe to rediscover the collective *I*, which was reasserted by the spontaneous organizations for self-defence formed by the working classes and which the bourgeois legislator was forced to swallow. In the stomach of the bourgeoisie, the bitter pill of the collective *I* was transformed and elevated, thanks to the legal socialists, into a justifiable object of scholarly enquiry, so that an 'affirmation of collective law' could now be made with some satisfaction.[56] This return of collective law represents no small development, given that the idea of the collective had been banished completely by the individualistic legal culture of the eighteenth and nineteenth centuries because it undermined the beautiful but artificial construction of the code.

The Origins of Comparative Law

The nineteenth century is the century of states, of laws and of codes. Its defining feature is the worship of these legal institutions. To the observer, the political and legal landscape of continental Europe in the nineteenth century appears as an archipelago of legal systems, each wedded to its own insularity. This impression is not softened by the frequent recourse to foreign models (the Napoleonic Code, for example) in the drafting of national legislation. Because of this insularity, the nineteenth century was not the best historical environment in which to foster comparative law: it is hard to come by the sort of cultural awareness that would see in comparison the sort of conceptual richness that might lead towards positive law.

In fact, it was the codes themselves that had become the representatives of a national positive law, pressing all the forces of legal culture in every country into service in their exegesis. Of course exegesis was an activity which, in and of itself, confirmed the written nature of the law: the law was thus completely confined within the bounds of the codified text, bounds that could not be exceeded because to do so would not help the jurist but would rather distract him. So long as the cult of the law and the jurist's self-definition as exegete persisted, there could be no space for comparative law. Such comparative efforts as there were, were the product of culturally ambitious individuals living in non-codified areas such as Britain (take Henry Sumner Maine), or Germany (Joseph Kohler). There was some activity in France, but it was almost entirely in the narrow field of comparative legislation. For example, in 1869 – the same year in which Henry Maine was appointed lecturer in historical and comparative jurisprudence at Oxford – the Société de législation comparée (Society for Comparative Legislation) and its accompanying journal were founded in Paris.

It would not be until the 1880s, as we have seen in our brief discussion of Saleilles and Gény, that the necessary openness of climate would prevail in order for comparative law to begin to be seen as an indispensable cultural necessity for any jurist, as an integral part of the habitual armoury of any lawyer, jurist or scholar, whose erudition surpassed the merely technical. It was thus in a context in which young jurists had begun to chafe under the yoke of statism and legalism, and the value of cultural pluralism as a means of taking account of diverse historical experiences had begun to be recognized, that comparative law emerged from its marginal position as the predilection of a few dilettantes and became its own autonomous branch of legal studies, with its own professional organizations supporting research and a proper place in the syllabus of the faculties of law.

It would be some time yet before comparative law arrived at the almost total ubiquity that it enjoys today. This was because legal positivism, that is to say the fallacious conviction that the law is nothing more than the orders made by the supreme political authority in a given territory, had placed a set of blinkers on the jurists working in Europe in the era of the codes, limiting their attention to the laws applying within precisely defined political frontiers. Nonetheless, we can date the origin of comparative law with some certainty to the provocative ideas of Saleilles and Gény.

In fact, it was Saleilles above all, with his intolerance for the status quo, his insights and his anxiety to rid French jurists of the weight of legal obscurantism, who forged the path towards comparative law. The great Burgundian civil lawyer found comparison very congenial to his aims – it was most certainly a necessity as far as he was concerned – and he dedicated all his energies to the discipline. He was certainly attracted by the gravitational pull of the common law, but also by the passage of the German Reich from a jurisprudential system (*Juristenrecht*) to a codified one. He followed the progress of the draft German codes with interest and was clearly fascinated by the high level of German scholarship that was being forged into a code. He communicated this complicated process to his fellow French jurists in an impressive series of works in which he placed this new German law in dialectical relation to its French equivalent. These included his essays on the general theory of obligations in the draft BGB, which were published in the *Bulletin* of the Society for Comparative Legislation in 1888 and 1889, followed by his study of the newly completed BGB in 1901 and his introduction to the study of German civil law from 1904.[57]

Saleilles did not stop there, however, and in 1900 he organized the first international conference on comparative law in Paris. Here he gave one of the foundational papers of the new field of study, which explored the concept of comparative law and what its aims should be.[58] It is in this work that one can see the best expression of Saleilles's far-sighted programme of comparativism, in which diverse experiences of the law are brought together in the hope of moving towards a unified body of law.

Saleilles was to continue contributing enthusiastically to the new field of comparative law through writing and teaching right up until his premature death. The last work he published in his lifetime was a set of twenty-five introductory lessons to a course of comparative civil law on legal persons.[59] But by now comparative law had its own, well-defined, shape: the Paris conference of 1900 had served as a solid foundation stone on which the edifice might be constructed.

Not long afterwards, Ernst Rabel (1874–1955), a Roman lawyer, legal historian and scholar of civil law, would provide two further foundations, one organizational, one theoretical: an institute for comparative law at the University of Munich, founded in 1916, and a very sophisticated methodological handbook to comparative law, in which Saleilles's insights were formed into a coherent intellectual programme.[60]

There is one final name I should mention: that of the Frenchman Édouard Lambert (1866–1947), one of the restless young jurists of the 1890s mentioned above (his doctoral thesis, examined in 1893, treated the controversial topic of contracts for the benefit of third parties). Lambert was appointed by Saleilles to the position of rapporteur-general for the 1900 conference. Thereafter he dedicated himself to establishing the field of comparative legal studies in France. In addition to his many scholarly contributions in the field, he was responsible for the creation of the Institute for Comparative Law at the University of Lyons in 1920, only the second of its kind in Europe after Rabel's Institute in Munich.[61]

Rabel and Lambert were also responsible for the first great comparative rapprochements with the universe of common law.

The Origins of Labour Law

The title of this section requires some explanation in order to avoid misunderstandings. The origins of what is now known as *labour law* are comparatively recent, even if individuals have been spending physical and intellectual energy to ensure their own subsistence since the dawn of time. But labour law requires a concept of *labour* – either under one's own command or in the employ of another physical or legal person – that is seen as a specific ethical, social, and therefore legal, phenomenon. This occurs only at the end of the nineteenth century.

The Napoleonic Code, slavishly followed by all of its successors, resolves the issue of labour, whether as an employed or a self-employed worker, by reference to the technical legal framework of rentals. This has the effect of stifling the issue. Let us dwell awhile on this point of great legal historical import, to allow the reader to come to a full understanding of the socio-political and legal effects of a choice that is only apparently a purely technical one. In fact this is a deliberate strategy on the part of bourgeois law, which is reconstructing an ancient legal framework from Roman law. One thousand eight hundred years of history appear to have passed in vain when we see the almost exact replication of provisions from Roman law in the section of the *Code civil* beginning at Article 1708.

Article 1708 begins by dealing with rental agreements. It states that 'there are two types of contracts of hire: those for things and those for work'.[62] The first is where the owner of a property, be it rural or urban, known as the lessor, offers enjoyment of that property to another individual (the lessee), in exchange for the payment of a fee. The second is a contract in which the owner of some labour offers the enjoyment of it to another party in exchange for the payment of a stipend. The only difference is that, in the second instance, the lessor, i.e. the worker, is the weaker of the two parties; the contract and the legal framework are the same. The labour which is the only property of a person who owns nothing tangible is itself simply a *thing*.

The strategy of the Romans, and of the modern bourgeoisie, underlying this unity between labour contracts and rental agreements is a shocking one: labour is seen entirely materialistically; it is commodified and reduced to a thing. There is thus a separation between the labour and the person of the worker, a separation whose main effect is to remove any ethical or social connotation from the contractual relationship. There is no such thing as a labour contract in the bourgeois codes; that noble aspect of the human being is buried, choked off and completely undervalued by comparing it to the rental of objects. If labour is just a thing, then it can be governed by the generic and all-embracing civil law, which enacts the cruel joke of elevating the worker to the status of property owner even if the only property he owns is his own energies.

Labour law as we understand it today – a substantially autonomous branch of the law – was thus impossible under such a regime. I should point out, moreover, that the perspective on this issue had not even changed by the time of the final code of the nineteenth century: the German BGB. There is no mention there of renting labour, but there is a mention of *Dienstvertrag* – a contract of service – which refers to a specific Germanic tradition dating back to the eighteenth-century Prussian ALR. The substance, however, is very similar to the French model, and the bourgeois underestimation of the realities of the labour market continues.

As we have seen in the previous few sections, there was a change of attitude and, at the same time, a more complex understanding of the problem began to arise. These changes were accompanied by two historical developments. The first development was the sometimes violent uprisings of the working class and the first concrete concessions made by the bourgeoisie to that class in the growing number of acts of 'social legislation'. The second was the intellectual revolution amongst young jurists, especially amongst those who took up a position of solidarity with the working class. Labour law was born in this environment, and hence was entirely separate from the purism and abstractions of the civil codes; it was the product of the rough, 'impure' and factual laws wrung out of the ruling bourgeoisie or paternalistically conceded by the more far-sighted governments, such as Bismarck's Germany. Labour law was born out of the daily practices of coalitions of workers who succeeded in making their voices heard. It was born out of the reflections of legal scholars, who managed to think up technical definitions for the new social advances. And so, labour law can be seen to emanate from a principally non-legislative environment, giving it a very specific shape, which it retained throughout the course of the twentieth century, despite repeated attempts to genericize it.

The two greatest contributions of legal solidarity were the affirmation of the unsuitability of the Roman concept of *rental of labour* to express the complexity and richness of the working relationship, and the assertion of a collective *I* (the worker as a member of an organized collective that gave his position greater weight and prestige), alongside the individual *I* (the cornerstone of the bourgeois system and the guarantor of its stability). This new idea of selfhood may not have created a new legal order, but it certainly added complexity to what was already present.

The refutation of the idea of rental of labour is very significant because it was that vision of the working relationship that had most served to denigrate the whole field of labour issues. Only once labour was liberated from the bonds of the decadent but efficiently conservative Roman-law tradition could it lay claim to the full extent of its value and overcome the humiliation it had suffered. The modern-day system of labour law could only begin to take shape once the carapace of rental of labour had been debated, its slyly retarding influence on the field had been recognized and it had finally been dismantled completely. All this is now well established, a situation which discredits the recent attempts in Italy to resuscitate the reputation of Lodovico Barassi (1873–1961), a civil lawyer steeped in the Roman tradition who, in his weighty study of 1901 entitled *Il contratto di lavoro* ('The Labour Contract'), continues unperturbed to assert the relevance of Roman legal categories for the newly begun twentieth century by identifying his subject as a purely rental-based relationship.[63]

The very early years of the twentieth century do see one step forward, however, in the shape of the German jurist Philipp Lotmar (1850–1922), a strong believer in solidarity, who after training as a Roman lawyer at various German universities became Professor of Roman Law at Berne in 1888.[64]

In 1902 he published the first volume of a very broad study on *Labour Contracts and Private Law in the German Empire*, which happily demonstrates a complex attitude towards its topic with the difficult aim of promoting diversity via continuity. Let me explain myself.

Lotmar unites and seeks to harmonize the conflicting instincts of a believer in solidarity, a Romanist and an acute observer of the surrounding realities of the German Reich, which now displayed the characteristics of an advanced industrial society. It is especially his capacity for observation – a novelty in a jurist – that causes him to assert the value of adding to the traditional sources of labour law provisions such as collective contracts, union regulations, arbitration panels including members who were not robed judges, workplace inspections, and surveys of workers. The resulting body of law is like a snake in the middle of trying to slough its skin which cannot completely manage to do so: the labour contract is still seen as part of private law and so fundamentally as an exchange agreement (work in exchange for pay) and yet the jurist expresses his sincere dissatisfaction with its categorization under the old and unsuitable headings of the traditional systems of rental and purchase and seeks sincerely for a new model.

Lotmar makes two points that should be emphasized. Firstly, labour contracts cannot be seen apart from the socio-economic situations in which they are intended to operate; indeed they must always be seen and interpreted in the context of what Lotmar calls the *faktische Umwelt* (the 'factual environment'). This is a very uncharacteristic observation for someone trained in pandectism to make. Secondly, labour contracts cannot be anonymous, nor can the work in question be reduced to a form of property. Labour is a person in action, and thus it is not only a form of property, it has an undeniably personal aspect to it; labour is an essential part of the life of the flesh-and-blood person who is the worker. And Lotmar concludes that it is this very close relationship to the worker's person that defines labour.[65]

There is one more aspect of Lotmar's work on which we should touch. Beginning in 1900, Lotmar worked steadily on the *Tarifverträge*, tariff contracts – embryonic versions of the collective contracts now very frequently negotiated by German trade unions.[66] Lotmar's perspective on these compacts is still exclusively that of a private lawyer: he sees them as exchange agreements and thus views the unions as equivalent to agents under private law, but it is striking how much attention he devotes to collective agreements, condemned by the pandectist tradition as purely sociological phenomena. Solidarity prevails over Roman and civil law, but only to a limited extent.

The division becomes starker in the work of another German jurist, Hugo Sinzheimer (1875–1945), whose writings all look towards the future, analysing the present in terms of acute and far-sighted predictions of what is to come.[67]

This division could come about thanks to Sinzheimer's peculiar training: he was free of any allegiance to pandectism, but instead had come of age in the culture of the German free-law movement and of Ehrlich's sociology of law, which sprang from the former's innovations. Sinzheimer wrought a substantial change on the methodology of the legal scholar. Free of the constraining boundaries of legalism and formalism (two attitudes which had always gone hand in hand), he refuted the identification of state and society and of law and statute. Instead Sinzheimer made reference to Ehrlich's notion of *living law*, that is to say law as expressed through the social forces beyond the apparatus of the state, which was to be considered equally as valid as the laws created by that apparatus.

The collective contract was a manifestation of the autonomy of social groups, which, as autonomous bodies, were capable of producing their own laws. The collective contract therefore had to be seen as a revolutionary new type of *Selbstgesetzgebung* ('self-regulation'). This in turn meant that labour law could not be seen as an individual right enclosed by the state in the cage of its codes, but rather a social law born of one of the autonomous groups that made up society.

Sinzheimer thus overturns Lotmar's position, seeing the collective contract as a *korporative Arbeitsnormenvertrag* (a 'corporate contract governing labour relations'). As he defines it in his revolutionary essay of 1907, it is 'an agreement between a professional association of workers, on the one hand, and an employer (or association of employers) on the other, which contains rules intended to govern the content of future, individual, labour contracts'. The collective contract thus becomes normative, the contract that has authority over the agreements with individual workers since it overwrites and replaces their individual will.

This was a scandal and a heresy as far as bourgeois legal individualism was concerned, but Sinzheimer was doing no more than recording the way the wind was blowing, and it was a wind that would continue to blow throughout the twentieth century, knocking down old structures and creating new paths forward.

Commercial Law between Legislation, Scholarship and Practice

We have seen that the nineteenth century was the century of the codes. Civil codes, first of all, because bourgeois culture needed an indestructible material on which to write its primer of fundamental laws: individual property; contracts (between individuals); and inheritance. But there were commercial codes as well and, as the century went on, the number of these legislative experiments increased. Commercial codes are connected to the interests of a professional class, unlike the abstract and wilfully immutable civil law system. This meant the codes that were inaugurated were immersed in economic facts and were exposed to their extremely rapid development. The commercial codes were legal material that had to respond to the underlying economic and technological changes of the times.

The *Code de commerce* of 1807, which was the third stage in Napoleon's imposing process of codification, is a very important piece of legislation. It certainly does not tear down the alliance between commerce and political power already established in Colbert's *ordonnance*: the mercantile world is considered part of the private sphere of the trading individual, even if the existence of the code means that the state is setting up the system of laws that govern trade. The code was drawn up by a committee of practitioners who followed what was by now an established tradition. It preserved a pre-existing matrix of rules that safeguarded the interests of the mercantile class, as well as maintaining in existence a special commercial tribunal composed of non-robed judges. But the code also makes a fundamental choice that demonstrates a further and perhaps overriding goal. The peculiarity of commercial law that sets it apart from the rest of the civil law does not reside in the distinctive characteristics of the individuals who practise commerce, as it did in the *ordonnance* of 1673, but rather in the specific facts of objective reality that are *acts of commerce*. The code seeks to safeguard and promote commerce as a vital aspect of the economy of the French empire.

There are a few notes to add: firstly, that the peculiarity of commercial law is deliberately understood in a restricted manner, in order to avoid impinging on the mother of laws, the Civil Code. The Commercial Code, coming some three years after the Civil Code in 1807, is careful not to let its particularism step out of its predecessor's shadow. Secondly, the Commercial Code was inaugurated at point before the industrial development of France had really begun; it therefore reflects a society that is largely still centred around real property and on a form of capitalism almost coextensive with trade. Thirdly, as the Commercial Code was being written, the jurists on the Conseil d'État struck out a number of the solutions suggested by the drafters on the basis of practical commercial experience, thus weakening considerably the code's claim to be a true reflection of the facts of contemporary economic life.

Despite these self-imposed limitations, the *Code de commerce* functioned as a model for the Spanish *Código de comercio* of 1829, the Portuguese *Código comercial* of 1833, the Commercial Code of the kingdom of Sardinia of 1842, and then its Italian counterpart in 1865. However, the *Code de commerce* was inevitably no more than a provisional model for these codes, since it dated extremely quickly. There were two reasons for this: the above-mentioned economic and technological development and accompanying large-scale industrialization which western Europe underwent from the 1840s on, and the breadth of vision of commercial law, a relic of its origins in the practices of the markets of medieval Europe, which meant that the protection of any one state inevitably chafed and that the solutions found for the same economic or legal problem in different political contexts were sure to be compared against one another.

The legislative product of 1807 was thus destined to decay very quickly, as were the other codes that modelled themselves on it. This decay caused a proliferation of special laws modifying the code or integrating new material vital to commercial organizations: negotiable instruments and documents of title; protection of patents; laws on banking and stock trading. Whenever there was innovation in economic life, a new special law had to be drawn up. These special laws not only constitute a preliminary foray into comparative legislation, since they frequently echo the solutions of foreign jurisdictions, but they also demonstrate the necessity for commercial law to respond primarily to the perceptible facts of the economy, rather than to the artificial projects of a sovereign. Two specific examples of this innovation were the cheque and the bill of exchange: the former an import from the English common law and the latter an abstraction made by German scholars and legislators that liberated the document of title from its specific factual bases, making it a sort of paper currency for the commercial sector.

One code that distinguishes itself from the others is the *Allgemeines Deutsches Handelsgesetzbuch* (ADHGB), the German General Commercial Code, inaugurated in 1861.

This document chooses to ground its specialism in its defining aspect of objective reality: the *act of commerce*. This, as we have seen, is no novelty. A unified civil code is still some way off in Germany, but this commercial code demonstrates the desire of the economically dominant class to foster and develop ever wider processes of political unification, since the wider compass of the unified political realm allows a space for the full expression of commercial traffic. Moreover, in the continued absence of a civil code (the consolidation that had taken place in Prussia in 1794 was by now hopelessly out of date), the ADHGB even encroaches on the civil law territory of contracts and bonds, and so constitutes a very broad effort at commercial codification. Given the constructive aptitude of German pandectists in the nineteenth century, parts of this code were adopted as emblematic, and the whole document was celebrated as the most theoretically and technically robust example of its kind. The ADHGB's openness was particularly lauded: it was seen as inspired by practice but not allowing itself to become enmeshed in the purely practical. For example, the ADHGB's establishment of the abstract nature of bills of exchange, mentioned above, is drawn from the great reservoir of pandectist concepts, which had already produced a law covering such documents in 1848.[68]

The Italian Commercial Code of 1882 demonstrates a more complex, more culturally florid legislative technique than its German counterpart. It too is founded on the objective notion of acts of commerce, and it also broadens its scope by legislating on issues that, whilst they originate in the mercantile sphere, are also used in other activities, such as insurance, public limited companies and bills of exchange.

Because of its constant attention to the practical, its ability to react to and incorporate innovation and change in economic life, and its inherent tendency to look beyond the borders of any one state, commercial law is always in the vanguard of legislative and scholarly progress. No one would have been happier to proclaim this fact than the greatest German commercial lawyer of the nineteenth century, Levin Goldschmidt (1829–97), who wrote as follows in his lively historical account of the field: 'general civil law cannot ever aspire to reach that level of liberty and mobility and that universal applicability that a law which must deal with the needs of commerce must necessarily maintain'.[69]

Although the connection to a socially powerful class is never lost, it must nonetheless be stressed that there is an awareness in these turn-of-the-century codes that the drafters are interpreting an objective order. This attitude is expressed well by a Spanish legislator in the introduction to the *Código de comercio* of 1885, which looks back on the final stage in a process of evolution that saw commercial law become 'an independent branch of law, with fixed principles derived from natural law and from the nature of mercantile activity'.[70]

An essential component of this development was also the scholarly tradition that had grown up around the commercial law and which rescued it from the pigeonhole to which it had been consigned by the crafters of the civil law by affirming its right to be considered a body of knowledge of equal weight and prestige. There are two main scholars working in this field whom we should mention, the latter the pupil of the former: the first was German, Levin Goldschmidt, the second Italian, Cesare Vivente (1855–1944). Both academics took the same methodological approach, devoting close attention both to the history of legal institutes as a good indicator of what was ephemeral and what was enduring, and to everyday economic practice as a fertile field in which innovation and change occurred. The historicity of law, so neglected by the pandectist civil lawyers in their pursuit of purity, was thus the bread and butter of the scholars of commercial law.

The First Codification of Canon Law in 1917

Earlier in this volume, I explained how the Roman Church, from the beginning of its history, sought to form itself into an autonomous legal entity which derived its authority not from any terrestrial power but from Christ as divine legislator. We have seen how this led the Church to set up its own, very particular, system of law – canon law – and we have followed the development of this branch of the law throughout the millennium of the Middle Ages.

During the modern era, the sources for canon law maintained their characteristic plurality, even though the shock of the Reformation drove the Church towards rigid centralization and hierarchy. Human canon law continued to demonstrate that elasticity that befitted its pastoral nature. The principle of *aequitas canonica* continued to dominate unchallenged, and indeed it could not be otherwise, since this principle was linked to the higher goal of spiritual salvation. Furthermore, the prominent role of judges and masters of laws as natural dispensers of equity continued, under the shadow of the supreme judge, legislator and governor – the Pope in his role as Vicar of Christ.

Because of this, unlike the situation we have observed in the states of continental Europe during the modern period, the popes never saw fit to draw canon law together into a complex of laws dictated from the centre. What the canon lawyers call *ius vetus* (the 'old law'), staggers on until 1917 as an enormous and venerable agglomeration of legislative, doctrinal and jurisprudential sources. Indeed, one could argue it was a chaotic morass not dissimilar from the medieval and post-medieval *ius commune*. The Church had no time for the codificatory mania, which smacked too much of the Enlightenment and of the revolution, both fierce antagonists of the Church and its policies.

There was some possibility of reform in 1870, at the First Vatican Council, a gathering that was cut short because of the conquest of Rome by Italian troops, but that nonetheless brought the benefit of collecting together in Rome representatives from the most remote parts of the globe. The perception of a need for a clearer and more certain model of canon law, in order to fit with the discipline's pastoral mission, began to circulate at the council. For the first time, the Church too began to consider seriously the idea of codification.

There commenced many years of sometimes acrimonious discussions. There were those who were hostile, sometimes extremely so, to anything that might compromise the elasticity of canon law, since they saw this as intimately connected to its pastoral mission. A code, with its emphasis on rigidity, would therefore have been contrary to that mission. But, in the end, the arguments in favour won out: on the one hand, there were the justifiable demands for greater certainty and clarity in the Church's rulemaking; on the other, a desire to imitate a movement that had swept across all the states of continental Europe and which now made the Church's refusal to adopt a code seem humiliating.

In 1904, Pope Pius X swept aside the delays and ordered that the work of codification begin. The committee was led by Pietro Gasparri: an excellent jurist and professor of canon law; he would later be better known as the Secretary of State to Benedict XV and Pius XI, in which role he was central to the negotiations with the then Kingdom of Italy that led to the Lateran Pacts of 1929. After a little more than a decade, in 1917, Benedict XI inaugurated the first *Codex iuris canonici* by means of the apostolic constitution *Providentissima Mater Ecclesia* (it is customary to call papal documents by the first few words of their Latin text). In the document the pontiff expresses his satisfaction at this latest product of what he liked to call the *ius ferendarum legum proprium ac nativum*, that is to say the way in which the 'native' law of the Church could produce its own rules – *native* because it did not belong to any earthly dominion but was rather originally of the Church.

This codification of the canon law has its merits and its defects – both those in favour and those hostile to it could find some substantiation for their arguments. There was, at last, a clear and certain text of the canon law that even the smallest parish would be able to assimilate; yet the canon law was most certainly made more rigid. The text certainly appears to have technical and legal rigour – a probable result of Gasparri's undoubted erudition; yet it also constitutes a sometimes excessive legalization of the sacred. For example, the traditional tripartite canon law division of the legal sphere into *persons*, *things* and *actions* becomes such a binding formula that the religious significance of the sacraments is effaced as they are forced into a category of *things* that seems ill equipped to hold them.

Without going into detail here, one question must be answered: leaving the obvious peculiarity of its content to one side, formally speaking, was the Church's codification merely a slavish imitation of its lay predecessors that abdicated the millennial traditions of canon law? In my opinion, the answer is that this is a legislative act that finds its own way of being a typical code. It certainly cleaves to the essential aspect of every modern code – exclusivity – and lays claim to be the only possible set of rules.

However, the Church's codification does not look upon the past with an attitude of smug superiority, as do all the other post-Enlightenment codes. The Church codifies its laws in a way that is in perfect continuity with its millennial history: a history which is considered to add richness to the present. Canon 6 (the Code uses the term *canon* in preference to *article*) expressly asserts the value of the legal learning of the past as an interpretative aid, inviting one to make use of previous canon law documents wherever there is no new provision in the code that overturns their substance.

Furthermore, it is instructive to note how, in Canon 20, the *Codex* confronts the inevitable problem faced by every code: the issue of gaps in a system of positive law. After the usual remedy of analogy, referring to other rules elsewhere in the code with the same or a similar object, the code leaves the legal horizon as broad as possible: the judge may have recourse to any of the general principles of law derived from canon law equity, from the practices of the Holy See, or from the consensus reached by the masters of law. Two other points worthy of note: firstly that the Church remains suspicious of any general principles that, considered in isolation, might not take fully account of the specific objective context of in which an action was committed. Such general principles must be reassessed in the light of canon law equity, that is to say, in the light of a thorough consideration of the particular in all its concrete detail. The second point to note is that the enlightening opinions of legal scholarship remain a precious source, meaning that this twentieth-century code continues to hand down the ancient legacy of the medieval *ius commune*.

3

Journeys in Contemporary Law

Twentieth-Century Law: The Crisis of the State and the Rediscovery of Complexity

The speaker at the opening ceremony for the academic year 1909/10 at the University of Pisa was a brilliant young Italian expert in public law called Santi Romano (1875–1947); he did not mince his words: his speech was entitled 'The Modern State in Crisis'. This was a statement of extraordinary frankness for someone who taught and researched in the areas of constitutional and administrative law. The young jurist had, however, observed for several years a process in which it seemed to him that the state was being 'eclipsed'. Romano saw this eclipse as closely linked to the state's inability to impose order on the increasingly complex socio-political and legal systems.[1]

Now that jurists such as Romano had begun to rid themselves of the distorting lens through which they had been looking at it, the fortress of bourgeois law began to seem like an impressive but artificial construction. As Romano puts it in his extremely lucid summary, the tableau had been reduced to two players: the macro-individual of the state, and the micro-individual of the single citizen. The sources of law had also been reduced: to the written laws of the state in the public sphere and to the contract in the private sphere. Society itself had been reduced to an anonymous mass of citizens, all formally equal, who submitted inertly and passively to the commands of the centre of power. In Romano's view, the ideas of the state and the individual had now been raised to a metaphysical level, leaving them floating isolated from one another and cut off from their moorings in society.

Romano's ringing condemnation of contemporary socio-political and legal thought ran counter to the reductionist trend of the previous century, which had sought to set up a system that was above all ordered and simple. Now, however, at the beginning of the new century, that system was plainly too ahistorical. Simplicity became oversimplification: an enforced lack of subtlety devoid of any link to reality and inimical to the real social, economic and legal developments that had grown ever faster over the course of the nineteenth century.

The social landscape of continental Europe had changed drastically. Not only was there the 'thriving movement of workers' organizations', hailed in 1894 by the Italian socialist Claudio Treves, there was a more general growth in all manner of spontaneously born associations, many of which were at odds with the plans of the exclusively bourgeois state.[2] Some of these associations were targeted with repression, some were grudgingly tolerated, but they were certainly a lively and effective presence in early twentieth-century Europe. In Romano's view, they were at the heart of the decrepitude of the great nation-building project that had begun with the French Revolution – a project that, it should be pointed out, Romano continued to admire. The crisis of the state lay in its loss of compactness, which could be attributed to the emergence of a collective *I* amongst the citizenry. The bourgeois power centres had always denied the idea of the collective *I*, because it represented the greatest danger to their projects: a collective *I* would muddy the simplicity of the bourgeois legal sphere by giving non-property-owning citizens a more substantial form of citizenship that was obtained via membership of a group.

Under the influence of this disruptive change, the two pillars of the post-revolutionary legal system – the stark separation between public law and private law, and the equally stark division between the world of law and the world of facts – began to crack.

Modern legal thought had recuperated the ancient Roman division between private and public, something that the Middle Ages had almost completely eliminated: the 'public' sphere was given over entirely to the state, and the 'private' dealt only with relationships between individuals. But the presence of a third dimension – the non-state collective – undermined this dichotomy that was so precious to the bourgeois order, which was founded not only on a deep-seated commitment to individualism, but also on an equally deep-seated commitment to statism in order to provide the necessary powerful protection to the individual property owner's interests.

The second pillar of the modern legal system – the rigid control of the legal sphere by the state and the equation of the law with the will of the state – also began to crack under the proliferation of economic and social phenomena that made new demands to which the citizenry responded without seeking the blessing of the powers that be. These were rough facts, but facts that communities could perceive, and they therefore laid the basis for an effective new body of law.

The final result was the loss of the state's authority and the collapse of its monopoly of the law. Two levels of legality thus developed: that of the written law, and that of everyday experience – formal law and living law.

Those in political power across all of Europe found themselves obliged, whether they liked it or not, to take account of the changed situation and to adapt themselves to it using the special laws we surveyed above. The facts that had emerged and been established through social practices were thus enshrined in legislation. At the same time, there was now a non-legislative means of generating law, leading to two levels of law-making – factual and legislative – that were not always in harmony. The clean lines of the old legal landscape were thus smudged and its artificial simplicity was replaced with renewed complexity.

Society had had its revenge on the state. The example of the nascent field of labour law was emblematic: it was born in the face of hostility, or at the most tolerance, from the bourgeois state, which understandably clung to the outmoded idea of the hire of labour. The field of law was created by the struggles of the trade unions and by the practice of collective negotiation, both of which had been made possible only shortly before by special laws grudgingly conceded by those in power.

At the apex of this metaphysical heaven, in which the person of the state and the abstract individual imagined by the pervasive mythology of natural law were set, was the code, whose articles now put forward a model of the law that was becoming ever more irrelevant to the crude facts of daily life. Despite its status as the universally applicable mother of all laws, each nation's code grew ever further from developments in society, whilst judges and lawyers, at sea on an ocean of experience, found it necessary to supplement the gaps in the words of the legislators. The number of special laws incorporated into the codes kept growing, further stressing their lack of vitality and separation from contemporary concerns. There emerged new sources of law that were based in a factuality that would have been unacceptable to believers in the pure law of pandectism and the codes, such as the decisions of the judges at arbitration tribunals and collective contracts.

The rediscovery of the complexity of the law allowed the complexity of the society it had begun again to reflect to win the day. The more prescient jurists did not delay in proclaiming this victory: it was again Santi Romano who, in 1918, the last year of the horrors of the First World War, drew together the conclusions from his speech at Pisa and from several other forays into the topic in a weighty essay called *L'ordinamento giuridico* ('The Legal Order'), that set out a robust general theory of law.[3] The book gives voice to the new awareness of twentieth-century jurists that the law belongs to society before it belongs to the state. The modern era had reduced the law by making it a province of the state and it was now the time to return it to society's embrace. The constraining legal monism of the age of codes was to be replaced by a pluralism that would allow society and law to live harmoniously side by side.

European Corporatisms

Corporatism is another term that is full of uncertainties and misunderstanding, and that therefore needs careful definition.

I shall use it here in its generic sense. Corporatism is an attitude of dissatisfaction with modern statism and individualism; it seeks to recover the complexity of the social and legal order. Within that order, corporatism seeks to bring the apparatus of power closer to society by privileging the role of all kinds of association or corporation as the means by which the abstract subject may be connected to the concrete network of social relationships.

In such a vision, the association or collective becomes the necessary protective shell for the individual subject, precisely because it rescues him or her from abstraction. The association or collective thus becomes necessary to the individual, especially if he is economically weak – if the individual *I* is in a position where he must suffer the violence of the strong, the collective *I* will effectively rescue that individual.

After the drastic rejection of corporatism during the Enlightenment and after the French Revolution had been incorporated into the bourgeois order, the second half of the nineteenth century was a time of renewal for collective movements on two levels: social practice and cultural projects. The number of associations grew during this period, as we have seen, beginning with the first workers' groups, which developed into political, charitable and co-operative organizations. These were a phenomenon that both united and distinguished the anonymous equal subjects of the bourgeois conception of citizenship; the centre of gravity of public life began to move towards these new social and economic associations. The associations took a great variety of forms: some, such as trade unions, fought to win emancipation for their members; some, such as professional organizations, sought to maintain privileges for their members; some sought to help their members and advance their interests, such as co-operatives; whilst still others brought together co-religionists, such as confraternities.

Alongside this spontaneous and irresistible outburst of change, which took place at the margins of an official culture that remained attached to its statist and individualist foundations, there was a continual series of changes in attitude that, in general, tended towards corporatism.

In Germany, the independent and strongly opinionated faction of Germanist jurists used corporatism as the standard to which to rally in their attacks against pandectism. We have already dealt with the Germanist Otto von Gierke's polemic against the pandectists' drafts of the *Bürgerliches Gesetzbuch* but, beyond the specifics of this particular debate, Germanistics was rising as a cultural current in its own right. In the mid-nineteenth century, when Georg von Beseler (1809–88), Gierke's mentor, set down the definitive Germanist's manifesto in his famous book *Volksrecht und Juristenrecht* ('The Law of the People and the Law of the Jurists'),[4] the distinction he drew was between the blindly individualistic Romanist tradition and its Germanist counterpart that placed great emphasis on the emerging associations. The mainstay of Germanistics was to be the *Genossenschaft* (the 'collective'). The *Genossenschaft* was also central, even titular, to Gierke's impressive reconstruction of an authentically Germanic tradition of law.[5] The Germanists' disagreement was not so much with the BGB itself, as with Gerber and Laband's invention of the state's legal personality, which took the subject in Roman private law as a model and therefore cancelled out the state's complex, communitarian identity, excluding its substratum of associations. In 1889 a pupil of Gierke, Hugo Preuss (1866–1925), who would later help design the institutions of the Weimar Republic, wrote a book that attempted to produce a concrete design for a state as community, with a distinctly corporatist structure.[6]

Also in Germany, Albert Schäffle (1831–1903), who was not a jurist but rather a sociologist trained in the increasingly important field of biological sciences, depicted society as a complex organism, whose building block is the social individual who may form load-bearing structures by joining with other individuals. For Schäffle, it was as impossible to separate the individual human subject from his social group as it was to separate the biological cell from the tissue of which it is part.[7] The French sociologist Émile Durkheim (1858–1917) took up Schäffle's key message whilst avoiding the suffocating biological idolatry of contemporary France: social groups are to be valued as an essential conduit between the masses and the political elite.

But corporatism would find other fertile terrain in which to put down roots: social solidarity based in religion. Already in the 1860s, an authoritative member of the German Catholic hierarchy, the bishop of Mainz Wilhelm Emmanuel von Ketteler (1811–77), was calling for such a move. Ketteler confronted directly the issue of the relationship between the 'question of the workers' and Christianity, proposing that individualism be rejected and the idea of the collective be placed at the heart of society. In Ketteler's view, the collective needed to be freed from all anti-historical nostalgia and adapted to the needs of mid-nineteenth-century industrial society.[8] Catholic thought welcomed the collective, since it provided a useful intermediary that might help society steer between the subversive desires of socialists and the conservative desires of the bourgeoisie. This welcome extended to the highest level, with the support of Pope Leo XIII's encyclical *Rerum novarum* of 1891, but it was also evident amongst those who were advocating a state founded on co-operation between the various parts of society. For example, Giuseppe Toniolo (1845–1918), an economist and sociologist with close connections to the Catholic hierarchy, was convinced that the collective was the solution to social conflict, yielding the positive result that the two opposite but equally serious risks of total state control and total anarchy might be avoided.[9]

The theoretical interpretations of the return of corporatism in the late nineteenth and early twentieth centuries thus presented a variety of attitudes: from naivety, to conservatism, to generous solidarity. In effect, corporatism served as an enormous empty vessel to be filled with a great variety of contents. It certainly helped bring together coalitions of the weak in the struggle for social equality – corporatism played a pivotal role in giving solid form to the very first trade unions. However, not long after that, in the 1920s and 1930s, corporatism would also serve as the socio-economic framework for some of continental Europe's totalitarian states, the reasons for which we shall examine below.

There is one collective phenomenon I should mention that could only be included in this section because of the very generic definition I gave of corporatism at the outset. That is the political party, which I cannot cover in detail here but which becomes an increasingly important entity during the twentieth century. During the nineteenth century, various factions, social associations and movements began to organize themselves as unions dedicated towards political action. These developed into fighting organizations that were dedicated to gaining power; their features were a strong hierarchy and a unified command structure that concentrated power in the hands of a leader. Galvanized by the broadening of the suffrage, which began in England and Wales with the Great Reform Act of 1832, political parties grew more tightly organized as time went on, permeating gradually down from the elite level until, by the end of the century, they could claim to be authentic mass membership organizations. Just as occurred with economic and political corporatism, the political party would also be appropriated, altered and mutilated by the totalitarian experimenters of the twentieth century.

The relevance of the widespread and extremely varied corporatist movement to our legal historical project is this: the number of actors on the European stage grew from the two solitary individuals of the state and the private citizen to now include the group. The new associations that sprang up allowed a collective *I* to come into existence alongside the individual *I*. Social and legal complexity was thus comprehensively recovered.

I cannot end this section without mentioning briefly a final phenomenon that is, in the broadest sense, a collective one: the business enterprise. Here too we find social combinations that tend at times to go beyond the simple relationship between an individual and a company. Business begins to be talked of in terms of enterprises or companies (*Unternehmen*) in early twentieth-century industrial Germany, where there was a need to find new models into which to organize economic life.[10] The business enterprise is, in fact, an organization that unites people, goods and services in pursuit of goals that are important on a socio-economic level. The enterprise is an organization within which the individuality of the employer and employee are attenuated, where property itself is yoked to a higher cause. For the first time, the law begins to talk in terms of *organizations*: an unwelcome term to modern economic individualism and its fixation with the private citizen and his individual relationships. An organization of course implies a necessary co-ordination of individuals and their relationships that may well enforce a heavy toll on that precious individuality.

The First World War and its Effects on the European Legal Order

The years before and after the beginning of the twentieth century are like a busy workshop whose courageous experimentations forged the first in a new set of legal tools, distinct from those of the old bourgeois order. These legal innovations were needed to govern a slew of developments – social, economic and technological. The beginnings of a new legal culture can thus be discerned in the turn-of-the-century era.

Amongst the many historical forces which combine to generate the change I have outlined above, one tragic event of colossal magnitude plays a decisive role: the First World War. The war sweeps over the entire legal order, changing it dramatically. Such a forceful and general assertion may seem strange, since it may not be obvious to the reader how a war, despite its obvious and far-reaching political ramifications, can affect any area of the law, other than maybe international law. Let us therefore examine this question in a little more detail, in order to remove any unwarranted confusion.

To start with, this was a war like no other: it was not a local skirmish to resolve a border dispute or conquer an extra slice of territory, but rather a shocking blow that shook the entire political establishment of Europe and many others beyond the bounds of that group. The war caused the states involved to relax their conservative influence on the old legal order and forced the acceptance at the official level of legislation of a number of practices that were already commonplace in everyday life. The jurist can thus distinguish a specific type of wartime legislation that has its own flavour. These are not sporadic legal acts but rather a part of a clearly defined series of large-scale legislative interventions often coordinated into an organic programme of works. (The exemplar here must be the programme enacted by the German empire.)

It may seem likely therefore that this wartime legislation was a series of interventions linked to an exceptional event and so destined to wither away after the event was over. But this was not what happened, and the more perceptive jurists were quick to note that this was the case. They examined these instances of new legislation and, instead of rejecting or ignoring them, they debated each act in the context of the main areas of legal reform, devising appropriate categories in which to situate their much-needed discussions.

The jurists were conscious of finding themselves on one side of a torn curtain, through the rent in which they could glimpse a new legal terrain that was in no way ephemeral. One noted Italian civil lawyer, Francesco Ferrara (1877–1941), an acute commentator on Italian wartime legislation, rightly asks if these acts constitute 'anomalies dictated by the needs and opportunities of the moment, or rather [...] the seeds of a new law that will germinate in the distant future'. His response is equally apt: 'during great crises [...] innovations arise and mature in the conflict, paving the way for a transformed and regenerated future'. He therefore sees the war as 'an occasion to accelerate and bring to maturity the developing legal reforms; a violent spur to evolution'.[11]

After this first necessary change in attitude had taken place, there was immediately another. Although the wartime legislation touched on a great number of issues, there were two that are particularly pertinent for our purposes here, because they went to the heart of the old bourgeois legal order that had been set up by the French Revolution and by the codifications of the nineteenth century. These were the question of legal sources and the area of civil law. Both of these were jealously guarded aspects of a legal system that was founded on a very tight control over the production of law and on a civil code that served in substance as a constitution. It should be no surprise, therefore, that in Germany, France and Italy it is the civil lawyers who rushed to debate these developments.

I shall limit myself to a brief summary of the most significant blows dealt to the previously untouchable certainties of bourgeois law. These appear to be more or less constant throughout the various acts of legislation produced by the states participating in the war.

The executive branch now became the default producer of all legislative or similar provisions, hence expropriating what was once the undisputed role of parliaments. Legislation nonetheless continued to be ratified formally by parliamentary act.

Arbitration panels were set up that could reach rapid and effective decisions based on equitable criteria that were left up to the judge's discretion. The hitherto immutable idea of *iudex sub lege* ('the judge under the law') was done away with.

The legal role of women was significantly strengthened, albeit with the excuse of urgency. The state also began to intervene at the very heart of the civil law. Property owners, for example, began to be subject to significant limitations on their liberty. Contracts were also reformed: debtors were to be forgiven the inability to service their loan in difficult times; the economically weaker party to a contract was given more protection (tenants, for example, benefited from a legal extension to all leases). These moves imposed limitations on contracts that were socially necessary in exceptional times, and yet they could not help but deal a mortal blow to the principle of the formal equality of the parties in a contract, which had been one of the unbreachable walls in the bourgeois legal fortress. Finally the reparations for damages suffered during the war were totally without precedent in the traditional conception of sovereign responsibility, since the damages in question were caused by a legitimate exercise of state power – namely war. The idea of sovereign responsibility – which before had been subordinate only to misconduct, not to legitimate behaviour – had therefore to be completely reviewed.

For now I shall pause here, because these observations seem to me sufficient to communicate how profound the effects of First World War were on the fabric of the law. I should conclude by noting that several of these legislative acts laid the foundations for future legal developments. The First World War can therefore be seen, from a legal historical point of view, as a significant staging point in the law's journey through the twentieth century. It is during the First World War that the future begins to make its presence felt since it is the event that exposes, makes official and legislates for the crisis that had for some time been brewing in the lower reaches of experience.

Weimar Communitarianism

The title of this section may need some explanation for the uninitiated reader. The noun adjunct *Weimar* refers to the city, capital of the former Grand Duchy of Saxe-Weimar, where Goethe and Schiller lived and died. It was here, in a conscious nod to these great figures of German culture, that the Constitutional Assembly of 1919 met to create a new republican state after the ruinous collapse of Kaiser Wilhelm II's empire. *Communitarianism*, meanwhile, refers to the nature of the new state's structure, and a very particular structure it was: one which, despite its mere fourteen-year duration, represents a social, legal and political experiment that it is worth our while to examine.

Staying in Germany of 1919 for a while: the immediate context was the recent and overwhelming tragedy of war, which left a widespread feeling of social discontent together with a disastrous economic crisis that would only be aggravated by the punishing reparations imposed by the Treaty of Versailles. The air was still thick with the possibility of sedition, which had not been completely driven away by the war, and from the east the country now found itself in the shadow of the Bolshevik revolution, so perfectly and bloodily carried out against the Russian tsars.

The Constitutional Assembly was heterogeneous, with many ideological disagreements; nonetheless, it managed to piece together a consensus. The liberal state was not to be resuscitated, but nor was the Russian model of the dictatorship of the proletariat to be followed. The resulting communitarianism was the product of a compromise between the various factions, although it was also the desired goal of the lively German corporatist movements, which were amply represented amongst the assembly. One peculiarity of the Weimar assembly was the decisive role which intellectuals, especially jurists, played alongside the politicians. The masterly and persuasive Germanistic vision of Otto von Gierke, who was living out the last years of his old age, hung over the assembly.[12] Unsurprisingly, therefore, two former pupils of Gierke, Hugo Sinzheimer and Hugo Preuss, were the attendees whose contributions most defined proceedings. Despite a number of differences of opinion, both men were committed and ultimately successful supporters of a communitarian state.

Let me develop a little more what I mean by this cryptic adjective, *communitarian*: it indicates a pluralistic conception of the state as a community of communities, the state more as *Volkstaat*, a state of the people, than as *Rechtsstaat*, a state of law on the liberal model. The *Volk* is not identified simply with an anonymous mass of citizens but with a very broad community of *socialized people*.[13] As the father of communitarianism, Gierke, had written some years previously in 1915, the state was to be seen, 'not so much as an association of individuals joined by a common destiny, as an association of individuals already joined together in groups, who are pursuing further and greater communal goals'.[14]

The Weimar democracy was a *collective* one, since associations played a special role in a state that was effectively a macro-collective itself.[15]

Collectives become important because the fundamental idea of collective democracy is the primacy of the interests of groups over those of single individuals. The state does not stand aloof from the actions of individuals, but is rather a body that sets out to intervene specifically in the mechanisms of production. It is this aspect of the Weimar Republic that makes it the perfect summation of the early twentieth-century state, demonstrating its separation from the legal absolutism and individualism so beloved of the French Revolution and of the bourgeois legal structures of the nineteenth century.

The republic created at Weimar represents an attempt to overcome the rigid binary of governors and governed, of representatives of imperial power and the subjects who were expected to obey that power. The foundation stone of natural law, which had seemed to bear much of the load of the bourgeois system, was now removed. The separation and even opposition between nature and history, which we have examined in detail above and which was so essential to bourgeois culture, now collapsed so that only history remained. The stale mythology of natural law was thus refuted; instead all now turned on the play of historical forces that breathed life into human communities, the state and the intermediate collectives.

The heated discussions in the Weimar Constitutional Assembly passed on a number of old concerns to the experimental republic they created. For example, although the citizen acquires new rights that are connected to the changed economic and social dynamic of the times – rights which do not abstract the individual, but leave him or her connected to a network of relationships – nonetheless the old individual rights of the liberal tradition are still asserted. For example the principle of private property is asserted even for the means of production, although a series of duties is imposed upon the owner of such means that takes the system a long way away from any kind of Lockean natural law.[16]

So there were numerous innovations in the Weimar Constitution, since the document sets out to express its own, postmodern, period, and it succeeds. Because of its self-conscious novelty, the text has the courage to set itself up as the figurehead for an entire nation. Nonetheless, as I have said, the forces of political conservatism present at the assembly ensured that there were numerous hangovers from the bourgeois modern period. The message of novelty was thus muddied and ambiguous, something that would eventually prove the death of the Weimar Republic.

The constitution certainly conserves an idea of liberty, which has been shaped by the new currents of the early twentieth century. Freedom no longer precedes the state as it did in natural law's artificial old lifejacket of the solitary individual. Instead, freedom is now a social good that resides in the citizen's relationship with others in specific communities and the populace at large. The linchpin of society is no longer property, nor is the fundamental subject the property owner; instead the cornerstone of society is work and the working subject. The most important area of law is thus labour law, a discipline to which the liberal constitutional thinkers did not attach any importance, indeed they could not because of their individualistic and proprietorial outlook. Amongst the new intermediate communities, immediately above the family, came the trade union. In Sinzheimer's ambitious vision the union was not a tool of subversion, but a rational partner of the state, endowed with equal powers and duties and capable of self-regulation within the dynamic system of collective contractual negotiations.[17] Finally there were the 'company councils', which were set up in Article 165 and constituted an embryonic form of industrial democracy.

But the Weimar Constitution's utopian project still had to confront the contradictions that stemmed from its ambiguity, just as the constitution itself had to gain the consensus of a divided parliament. Despite the proclamations in favour of labour law, the document gave no firm indications as to how its effectiveness was to be guaranteed. Unions thus became mere statutory corporations (*Körperschaft*), beholden to the state or the community they served, whilst the company councils remained nebulous entities in the face of the private ownership of the means of production.

To conclude this brief summary of the Weimar legal system I should underline one point that draws together and develops what I have said so far. The Weimar Constitution of 1919 begins a new phase in the history of constitutional law, since it no longer casts itself as a *bill of rights*, that is as a philosophical and political catalogue that lists situations in which the state undertakes to respect the generic citizen's rights, but rather as a true *constitution*, a legal framework that governs and reflects the complexities of a nation's society. The Weimar Constitution sets out be a faithful mirror of German society and an interpretation of the country's historic values, which it seeks to translate into rules and principles for daily life.

The assembly at Weimar was the first time a state had been conceived of as serving more than one social class; it was therefore the first time that the full complexity of society was considered in a constitutional setting. Because of its open and completely non-classist consideration of society, the Weimar Constitution also marked the first time that the private citizen was seen as a concrete, not an abstract, being. The citizen of the Weimar Republic was understood through his experience. The constitution therefore spends more time considering factual social inequalities than it does the formal and abstract equalities of the 1789 revolution.

Finally, let us briefly glance at the interior structure of the constitution, because this is innovative in itself. After a preamble dedicated to the 'Structure and Tasks of the Republic' (*Aufbau und Aufgaben des Reichs*), the second part of the document seeks to enumerate the 'Fundamental Rights and Duties of the German People' (*Grundrechte und Grundpflichten der Deutschen*). This second part is divided into headings: the first treats 'The Individual' (*Die Einzelperson*); the second 'Community Life' (*Das Gemeinschaftsleben*); the third 'Religion and Religious Associations' (*Religion und Religionsgesellschaften*); the fourth 'Education and Schools' (*Bildung und Schule*); and the fifth 'Economic Life' (*Das Wirtschaftsleben*). Even this fleeting list of divisions should be enough for to the reader to grasp the size of the gulf that separates the Weimar experiment from the bills of rights of early constitutional law.

The Soviet Union and the Construction of a Socialist Legal System

This book has so far neglected to concentrate on Russia, a place that, during the entirety of the medieval and modern periods, remained aloof and isolated on Europe's easternmost fringe. At first Russia was a tributary of Byzantine culture to its south; later it assumed its own identity, very separate from that of the west, which consisted in a symbiosis between the Orthodox Church and the Russian state. There was some penetration by Enlightenment ideas in the second half of the eighteenth century, during the reign of Catherine II, a German princess who became tsarina. This trend was only a parenthesis, however, and for all the rebellions and pseudo-reforms of the early twentieth century, the Russian empire was still a largely intact Eastern autocracy on the eve of the revolutions of 1917.

At the end of the tsarist period, the legal order in Russia consisted of two levels: a popular law based in custom, which was unofficially tolerated by the regime, and an official body of legislation, which had been brought together in the early nineteenth century in the so-called Corpus of Laws (*Svod Zakonov Rossiiskoi Imperii*). This was an enormous compilation of legislative acts, whose tenth volume was dedicated to civil law. The Corpus of Laws was, in my view, a consolidation rather than a codification: it lacked the rationality, systematization and insistence on exclusivity typical of a true code. It is salient to note that the constitution's main drafter, Mikhail Mikhailovich Speransky, a minister in the governments of Alexander I and Nicholas I, was fond of emphasizing how authentically Russian the Corpus was: 'our legislation has found its sources in itself'.[18]

There followed the radical, destructive and tempestuous event of 1917's second revolution: the October Revolution that swept the Bolsheviks to power. This was an especially radical break because the new masters of Russia aimed to overturn completely the old socio-political order and to put in its place a new one that stemmed not from political struggle or protests in the street, but from a new vision of history and society: that of Karl Marx, who claimed his work was based in scientifically verifiable analysis. There certainly were struggles and protests, some of them very bloody, but their extraordinary momentum came from a philosophical position that had become first an ideology and then a programme for a civilization.

The prime mover behind this radical subversion of the Russian status quo was Vladimir Ilich Lenin (1870–1924), who was not a philosopher himself, although he was philosophically trained; instead he was a formidable man of deeds, who was able to transform a philosophical message into a plan for political and social organization. *The State and Revolution*, written in summer 1917, immediately before the revolution, served as the construction manual for the new state mechanisms that would replace the useless fragments of the tsarist political system.[19] The goal of the revolution was to install the *dictatorship of the proletariat*, which Lenin saw as the most democratic solution that would safeguard the interests of the greatest proportion of the population. The proletarian state would be intrinsically democratic because all of its forces would be dedicated to serving the proletariat.

The heart of the revolutionary project was the party and one party alone: the Communist Party. The party was understood by Lenin as the proletariat's class-consciousness given political shape – it was an all-embracing reality in which individual and mass might blend into one organic unity. The party had a precise and urgent goal: to create a state in its own image. One might even say: to make itself a state. The party is not identified completely with the state; rather, the state is the irreplaceable instrument of its will. The state is the apparatus of power that has the coercive ability the party needs to achieve its aims.

The Soviet plan for the rebirth of Russian society therefore took on the following form: a single, mass-membership party; a single ideology that aimed for social transformation; the wholesale politicization of social relationships; the close control of individual behaviour; the subjugation of the state to the party and the instrumentalization of the state to the Marxist-Leninist ideology represented by the party; and the subordination of the legal system to the revolutionary politics of the party.

Here then is a brief summary of the principal developments in legal history in the new Soviet era.

The first years of the new republic (1918–21) are the years of what is called 'wartime Communism': the law consists in decisions made by the revolutionary party, which take the form of decrees from state bodies or decisions of the supreme committees of the party itself. Alongside this runs the free creation of law by non-jurist judges, who were chosen for their proven faith in the revolution and thus exhibited a revolutionary legal conscience.

The years that followed (1922–8) were the time of the so-called New Economic Policy (NEP). It was decided, in 1923, to suspend the continuing revolutionary activity and to seek a compromise between the state's economic policy and private economic interests. The body of revolutionary legal thought that had been building was demoted to auxiliary status and the various republics of the USSR undertook processes of codification, beginning with the Civil Code of the Russian SSR in 1923. It should be stressed that these codes still bore the hallmark of the bourgeois codes of the nineteenth century.

Finally, during the years of the Stalinist dictatorship (1929–43), there was a demand for a more compact form of statute law and a return to a unitary and positive vision of Soviet law – in effect an increased statism. The jurist who helped the dictator achieve this tightening of the state's grip on the law during the 1930s was Andrei Vyshinsky (1883–1954), who would later become better known by a wider public as the USSR's minister for foreign affairs. Vyshinsky reduced the law to a series of binding statutes, whose force stems from the absolute primacy of the political will which the law is meant to serve. The law thus becomes the express will of the party made manifest through the legislation passed by the Supreme Soviet, or the Federal Parliament. Ever the propagandist, Vyshinsky frequently stresses the originality of Soviet law, thus repeating the similar proclamations of the tsarist minister Speransky, made more than a hundred years before. In fact, Vyshinsky's work makes widespread use of intellectual models, logical categories and technical solutions drawn directly from the Western tradition.

Immediately after the end of the Second World War, which brought many of the states of central and eastern Europe under the political and economic influence of the USSR, the legal model of the Soviet state was exported to these satellite states more or less unchanged. This created a region with a unified legal system which scholars of comparative law came to recognize as a system that ranked alongside civil law and common law as one of the primary legal systems of the world, possessed of its own distinguishing characteristics. However, since the beginning of the 1990s that chapter of contemporary legal history has closed, with the restoration of multi-party parliamentary democracy to central and eastern Europe and the consequent eradication of the Soviet legal system.

The Law in the Totalitarian Regimes of Italy and Germany

The twentieth century was an era of totalitarianism. After the early Russian experiment in setting up the dictatorship of the proletariat along Marxist-Leninist lines, other totalitarian states followed in Europe. These were of a different nature but still comparable to the USSR, since they resulted in a totalitarian political and legal sphere with the same fundamental characteristics noted in the previous section: the imposition of a single ideology; the existence of a single mass-membership party; the defining role of the political leader; a complete politicization of social relationships; and close control over individual behaviour. Here we shall examine the Fascist regime in Italy and its National Socialist counterpart in Germany. This is partly because of the historical importance of these two regimes, but also because totalitarian governments in other countries (for example in Franco-era Spain or Salazar-era Portugal), tended to follow the Italian model.

Fascism sought to replace the anarchy and anachronistic individualism of the liberal regime, which hoped at all costs to avoid submitting to a Communist dictatorship like that of Soviet Russia. Fascism's solution was to set up – with the support of the economically powerful sectors of society – an authoritarian state. However, the Fascists gained power by abiding within the letter of the constitution and they were always careful to keep up the mask of formal respect for tradition. In formal terms the Fascist legal system continued to plough the furrow of the liberal system it replaced, at least insofar as the primacy of the written law was consistently reasserted. However, in substantive terms, the former plurality of political parties was brought to an end. In its stead, a far-reaching system for organization and control of the populace was set up, leading to the creation of institutions that were clearly harmful to the rights of citizens, such as the Special Tribunal for Defence of the State.

The single-party state is a radical novelty in Italian terms as well. The National Fascist Party acted as a necessary conduit between private citizens and the state – the privileged place for the individual's political activities and engagement. Thanks to its paramilitary organizations, its various mass membership groups and its after-work clubs, the party was able to able to discipline Italian society, train its future leaders and fashion itself as the representative of the Italian nation. The state was to achieve its aim of safeguarding the national interest through the party and the party alone. It is unsurprising therefore that that state grew consistently larger over the twenty years of the Fascist regime. The two important dates in this growth of the state were 1928, when the supreme body of the party, the Grand

Council of Fascism, was 'constitutionalized' and the functions of head of the party and head of government were fused; and 1939, when the Chamber of Deputies (the lower house of the Italian parliament) was replaced with the Chamber of Fasces and Corporations, eliminating all relics of the old parliamentary regime.

In the field of law, Fascism worshipped the written law in all cases, even when sometimes the content of its legislation was abhorrent, as in the so-called racial laws of 1938. The regime gloried in its role as legislator, setting up a number of courageous projects. It also made a series of public interventions in the economy, such as the *bonifica integrale*, a programme of land reclamation on a massive scale. But above all the Fascists stood for wholesale codification: the Penal Code and Code of Penal Procedure were both published in 1931; the Civil Code and the Codes of Civil Procedure and Shipping followed in 1942. These documents survived the fall of Mussolini and, although they have been liberated of a large amount of Fascist baggage, they remain at least partially in force in Italy today. The Civil Code that entered into force in 1942, a unified code of private law that incorporates the ground covered by the former Commercial Code, was drafted by a commission made up of the most illustrious Italian jurists of the time. It incorporates the most significant advances made in legal scholarship during the first half of the twentieth century and represents a great step forward with respect to the old French-influenced document of 1865. The fifth book (*libro*) of the code, entitled *Del lavoro* ('On Labour'), is of particular interest. The book appears on the one hand dominated by a newfound attention to the idea of labour, and, on the other hand, fascinated by the organizational unit of the enterprise, which was, as we saw, a body favoured by the more innovative participants in twentieth-century commerce.

One phenomenon that should be noted is that, as well as the usual sycophants in and around the regime, there was also a community of very able legal scholars, devoted to the rigour of their discipline. A striking example of this came at a conference called in 1940 to discuss a proposal by the Fascist minister of justice to include in the draft Civil Code and submit to the Grand Council of Fascism for approval a preamble setting out the general principles of Fascist justice. However, the jurists had the courage to live up to their most significant role in society and reject this proposal, perceiving that the overtly political preamble with its clear link to Fascist ideology would introduce material that was alien to a document meant to serve as a rulebook for everyday life in Italy. The system of laws set up by the Fascist-era codes, especially the Civil Code, the Code of Civil Procedure and the Shipping Code, is founded in legal scholarship of the highest calibre.

There was one choice made by the Fascist state that should be emphasized because it represents one of the defining features of twentieth-century Italian legal history: the setting up of a corporatist sector within society and the economy. This sector rejected the energies of solitary individuals and looked to organized collectives and the interests which these brought with them, interests that were then reconciled with the overarching national interest. On a socio-political level, corporatism helped to overcome the dualism between society and state; on an economic level it permitted cooperation between the various categories of production.

The Fascist programme had been clear from the outset, declaring war on two enemies: individualistic liberalism and collectivist Bolshevism. The choice of corporatism, which the regime would make official via a complex piece of legislation in 1926, was thus a necessity and was made without enthusiasm. The nature of the corporatist sector, even such a corralled and subjugated one as existed under Fascism, means that it will always represent a crack in the compact social and economic structure desired by totalitarianism.

And corralled it certainly was. On a political level, the first and most confining influence was the party, the single party of Fascism. On a socio-economic level, the constraints came from the Act of 1926, which granted legal recognition to workers' and employers' associations (and, in so doing, abolished trade unions at a stroke), declared strikes and lock-outs illegal, and set up a special tribunal system to deal with workplace discord. The corporatist sector was thus covered by an authoritarian carapace that most certainly constrained its free movement, reining in the pluralism that is the defining attribute of any corporatist system.

The regime's unavoidable progress towards corporatism was thus filled with contradictions and the new order was constructed only slowly, remaining still incomplete when Fascism collapsed in 1943. However, the ambiguous corporatism of the 1920s and 1930s did permit the curious and perceptive Italian jurists to observe and understand things that would be essential to the future development of law in Italy after Fascism. The law of the totalitarian states underwent the same fate as the exceptional wartime legislation of the First World War: despite various distortions, which were soon identified and removed, it contained a number of insights into the contemporary historical period and presentiments of future developments. A cursory reading of the proceedings of learned conferences or the scholarly journals of the 1930s will demonstrate as much.

The totalitarian National Socialist (Nazi) state in Germany began its reforms of German society and state on 30 January 1933 when Adolf Hitler ascended to the chancellorship. National Socialism constituted a more radical experiment than Italian Fascism because it made a complete break with the preceding forms of state and sought to put down new roots.

As with Fascism in Italy, the motivations behind the upheaval in Germany were the social and political conflict and the economic crisis that had beset the Weimar Republic. Again the new regime came to power by constitutional means. But the subversion of democratic values was immediate in Germany – it was consummated mere months after Hitler's ascension by a series of measures: the suspension of fundamental rights; the granting of power to the executive to pass ordinary and constitutional laws; the elimination of political parties other than the National Socialist Party; the abolition of the independence of the *Länder* (the regions into which the German Reich was divided); and the setting up of a Special Tribunal for political rights.

The new roots that I mentioned just now can be summarized as an essentially racist programme, since racism was the central tenet of National Socialism. Italian Fascism was not built on racism to the same extent as Nazism; only after Italy's colonial expansion into Ethiopia and the establishment of the Empire in 1936 did Fascism start to incorporate racist elements into its outlook. But racism had been part of the original message of Hitler ever since his seminal book *Mein Kampf* ('My Struggle'), written between 1923 and 1926, which had put forward such ideas as history as a conflict between superior and inferior races; the Jewish people as parasitic leeches; the identity of a people as being defined racially; the identity of an individual as stemming from membership of a people; and the state as a means to safeguard the people's racial purity.

At the heart of the new Nazi vocabulary was the *people*, the *Volk* – a notion now very different from that invoked by the eighteenth- and nineteenth-century Romantics and historians. This radical new *Volk* gained its identity from its supposed biological unity: the *Volk* was a people which shared blood ties, thanks to which it could become a *Gemeinschaft*, a spiritually and socially united community. The new *Volk*, purified of the degradations of history, could rediscover its lost identity and its function as a *Volksgemeinschaft*, an extremely coherent and united community.

The essential function of the party was therefore to be the interpreter and concrete incarnation of the *Volk*. It followed that there could be but one party: the Nazi Party. The party therefore assumed the role of founder of the state; its responsibility was to transform the state into an apparatus of power congenial to the expression of the popular voice. State and party became inseparable, and the necessary linchpin holding them together was provided by the head of state, who was also head of the party: the Führer. The Führer was able to lead both institutions because he was the only person able to express the totality of the people's will in all its profundity. In Nazi ideology, the Führer is the personification of the people. He is the leader: a figure of extraordinary charisma and creative initiative, in whose hands all powers must converge. The Nazi state is, therefore, the perfect antithesis of the liberal model: it is a total state, *Völkischer Führerstaat*, that expresses the organic community of the people in the person of the supreme leader.

And what of the law? Understandably, in this new order, the legal structures also had to be built from scratch. The law was assimilated into the values of the *Volk* and the judge assumed a leading role. Unlike the judges in the liberal tradition, who were armed only with formal logic, the Nazi judge's authority stemmed from his role as natural interpreter of the values of the people, whose verdicts would give those values concrete application.

A new type of jurist was also needed, therefore – one who was unharnessed from the old bourgeois legalism and formalism and was, instead, immersed in the National Socialist vision of the daily lives of the *Volk*. As early as 1933, an Akademie für Deutsches Recht (Academy of German Law) was set up in Munich, under the leadership of Reichsrechtsführer (National Legal Officer) Hans Frank (1900–46), and the construction of the new legal system began. Many of these jurists showed a large degree of servility towards those in political power, and even bent the practice of legal scholarship to breaking point in order to provide 'academic' legitimacy to the horror of a system of law based in racism. However, the intellectual quality of some of the scholars did lead to some revisions of the bourgeois civil law that would later be of benefit to German private law scholarship after the fall of Nazism.

The old BGB, which breathed the rarefied air of an abstract and pure conception of the study of law, was of course found insufficient to the task of governing the new German society. It was soon decided that it should be replaced by the *Volksgesetzbuch*, the 'People's Code', which was to be drafted by the Akademie in Munich. The war prevented the implementation of the new code, although it was completed in 1938.

The final point to note in this thumbnail of National Socialism is that there is no room for any type of continuation of the Weimar Republic's *collective democracy*. The Nazi era bears witness to a progressive dismantling of trade unions and company councils, and labour law thus came under the sway of a rigid statism. The *Führerprinzip*, that is to say the general principle that all power is handed to a specific leader, was applied even within businesses. Every enterprise was thus constituted as a community defined by a leader (the entrepreneur) whose subordinates were expected to show complete obedience. The entrepreneur himself was then subject to the various national directives, to follow the approved formulae for contracts determined by the Ministry of Labour, and to comply with ministerial ordinances.

After 1945, New Constitutions and Old Codes: The Beginnings of European Unification

In the years 1939–45 Europe and the entire world suffered the horror of a long and devastating war. The Second World War profoundly affected the structure of European law because of the new social, political and economic changes it wrought.

In Italy and Germany the totalitarian Fascist and Nazi regimes were overthrown, although the Fascism of Franco and Salazar would live on for some time yet in Spain and Portugal respectively.

The Soviet Union persisted in its faith to Marxist-Leninist ideology, with one significant innovation: those countries of central and eastern Europe that were brought under the political sway of the USSR had the Soviet model of political and legal systems imposed upon them. This political unit, which was kept starkly separated from western Europe by the so-called Cold War, becomes an autonomous legal zone ruled by what comparative lawyers call 'Soviet Bloc law', until the events of the late 1980s. With the sole exception of Yugoslavia, which created its own brand of socialism, the German Democratic Republic (DDR), Poland, Czechoslovakia, Hungary, Romania, Bulgaria and Albania all shared in a system, with few substantive differences between them: the dictatorship of the proletariat; the Communist Party as single party; state ownership of the means of production; and the total enclosure of economic life within the centralized apparatus of the state.

In western Europe there are two main points to mention: the process of constitutional renewal in Germany and in Italy, and the continued use of the old codes.

France also rewrote its constitution in 1946, enhancing the role of its parliament. But there is substantial continuity with the previous tradition of parliamentary democracy in the country, which was not abrogated totally as it was elsewhere (if one discounts the partial exception of the short-lived Vichy puppet state). It was in Germany and Italy, where totalitarianism had been torn down, that there was a need for a totally refounded state.

Germany was now a completely divided country, with her eastern portion coming within the Soviet sphere of influence; we shall concentrate here on the constitution of West Germany.[20] The West German Constitution is a very peculiar document, something which stems from the lengthy military occupation of the country's territory by the armies of the victorious powers in the Second World War. In 1948 the military governors of Germany encouraged the governments of the now reconstituted *Länder* to begin the process of drafting a constitution. Given the permanent division of the German nation's people and territory, the proceedings were to have a provisional tone: there was no constitutional assembly, merely a Parliamentary Council (the Parlamentarischer Rat), whose task was to draw up, not a constitution (*Verfassung*), but a 'Fundamental Law' (*Grundgesetz*), which entered into force on 23 May 1949. As Article 20 proclaims, the document set up a 'federal, democratic and social state', in which the people were to exercise their sovereignty via free elections in the established Western parliamentary tradition. The Parliamentary Council took heed of the weaknesses that had brought an end to the Weimar experiment and established the traditional rights of individual liberty not as overarching principles (as the Weimar Constitution had cast them) but as immediately enforceable laws. They then sought to defend the *Grundgesetz* itself by writing in Article 18 that 'anyone who abuses their right to freedom of expression, in particular the rights to freedom of the press, to freedom to teach or to freedom of assembly, in order to strike at the foundations of this democratic and liberal system will lose those rights', and by defining as unconstitutional in Article 21 any political party that seeks to 'damage or destroy the democratic and liberal order'. A Federal Constitutional Court (Bundesverfassungsgericht) was set up to act as the ultimate guardian of the constitution.

In Italy, after a referendum that abolished the institution of the monarchy, a Constitutional Assembly began to draft the constitution that would eventually enter into force on 1 January 1948. This document can clearly be seen to be the combined effort of the three groups represented in the Constitutional Assembly: the Catholics, the liberals and the Marxists. The Italian Constitution is not a bill of rights but a true constitution in the Weimar mould. It not only deals with the 'Structure of the Republic', to which the second part is dedicated, but begins by establishing some 'Fundamental Principles' (Articles 1–12), followed by a first part that analyses in detail the 'Rights and Duties of Citizens' (Articles 13–54). In the opening articles, the sovereignty of the people is established; the inalienable rights of mankind are recognized and guaranteed; the role of social groups is recognized; there is an undertaking to equality on levels other than merely the formal; and the unity and indivisibility of the Italian Republic and its constituent regions are asserted. Significantly, a judge is set over the written law: the Constitutional Court, whose job is to protect the constitution.

The title of this section talked of new constitutions and old codes. And it is true: the long life of the *Code Napoléon* continued in France, and the BGB remained in force in West Germany, for the Nazi People's Code had never progressed beyond the draft stage. The *Code Napoléon* and the BGB represented two models of liberal codification, so it is unsurprising that they were allowed to remain. The continuity in Italy, however, is more surprising, since the Italian Civil Code had been planned and drafted entirely during the Fascist era. We have already seen how, apart from certain forced insertions by the Fascists, the document was substantially the work of jurists – some judges but mostly scholars. These jurists had had the great merit of interrogating and interpreting their own historical context in a way the regime, with its foolish pretensions, could not. This conclusion holds for all the Italian codes, although the hand of the authoritarian regime is more clearly present in the penal codes of 1930–1, as one might expect given the close connection between punishment and public order. In the period immediately after the war there was some debate in Italy about what to do with the codes, but the more conservative position was clearly the majority view and accordingly the codes were subjected to a 'disinfestation', which removed the Fascist 'parasites' nesting in the folds of the documents (the image is not my own, but rather that of the great Italian procedural lawyer, Piero Calamandrei).

I cannot finish a book on the history of European law without at least mentioning the grandiose and tortuous process, which continues to this day, of forging a legally and politically united Europe. I call it grandiose because the edifice under construction is an imposing one, which has gathered ever more states within its compass. It is tortuous however because, in order to achieve the goal of a legally and politically unified continent, one must somehow bridge the gaps between the archipelago of national islands that make it up. The major developments of the European project have been as follows: the Treaty of Paris in 1951 instituted an embryonic form of union – the European Coal and Steel Community (ECSC); the Treaty of Rome in 1957 then created the European Economic Community (EEC), as well as the European Atomic Energy Community (Euratom); the Maastricht Treaty in 1992 gave birth to the European Union, now a body concerned not only with coal, steel, the economy and nuclear energy but also with foreign policy, collective security, justice and home affairs. Recently a Charter of Fundamental Rights of the European Union was drawn up, often called the Nice Charter, because it was signed by EU leaders at Nice on 7 December 2000, having been earlier approved by the European Parliament and Commission. This charter was an integral part of the treaty for an EU Constitution, put forward in 2004, but this project has not been brought into force because of the refusal by France and the Netherlands to ratify the document.

From the Old Millennium to the New: Some Final Considerations

As a historian, I believe in maintaining an in-depth knowledge of the present in which I live and work in order to give my view of the past the necessary acuteness. I hope therefore, that I may be permitted a few final observations on the journey of the law in these past few years before and after the turn of the millennium. Today, in 2006, the law and the jurist (that is those who dedicate themselves to the discipline of law, whether as practitioners or as scholars) are living through a moment of great uncertainty given the rapid and sweeping changes that are occurring. This is not an easy moment to be involved with the law, but, from a historian's point of view, it offers great potential. This is because a new legal landscape is being wrought and its features have yet to be defined because of the difficulty of the task at hand. The clear and simple legal landscape of yesteryear is being left behind, as it should be: it was too clear and too simple to reflect the underlying social reality in all its complexity. The idols at whose feet the old legal mythology of the modern period worshipped – the law as a product of the state, the written law, the principle of rigid legality, the principle of rigid separation of powers, the hierarchy of legal sources – all appear to have been largely torn down.

The two spheres of civil law and common law have recently ceased to be so starkly divided and have begun to intermingle. Moreover, because of its institutions and the scale of its production of laws, the European Union – with the United Kingdom as an essential part – has become the forum for even greater fusion. On the continent, the state, which had been the undisputed master of the legal sphere in the modern period, has consistently demonstrated itself to be incapable of providing society with its legal order and the written law is beginning to come down from its pedestal of the essential instrument for producing law. Instead, we are seeing other powers, particularly economic ones, begin to challenge the hegemony of politics and start to construct new and more congenial legal institutes. At the same time, globalization is expanding people's gaze beyond national boundaries, something which often reveals those frontiers' arbitrary nature.

Alongside the official forms of law there are now various other laws stemming from a series of alternative sources. The new law is thus often shaped during the process of legal practice, the most unforgiving workshop of all. The determining criterion of legality used to be that of validity, whether a decision or a rule corresponded to a generally authoritative model (usually that set up by the state); today the criterion has become effectiveness, whether a rule or a legal institute can reasonably be observed or applied. Readers who have been following the argument of this book attentively may be tempted to believe that the medieval legal cultures described towards its beginning are undergoing a renaissance. This is a useful way to look at the contemporary legal landscape, and yet somewhat misleading: it demonstrates the relativity of legal innovations that were thought to represent the last word in progress; but it runs the risk of obscuring the identity of our postmodern moment which has been shaped by forces born in the here and now and that must therefore be ordered according to the demands of the here and now.

But I would not want my reader to be afraid of the shifting ground under his or her feet. The movement is caused by excavations in which the foundations of a new legal edifice are being laid, a construction which will be in harmony with the requirements of these new times that have been maturing over many long years.

Further Reading

Preliminaries

Europe

Chabod 1964
Braudel 1987

The law

Grossi 2003a, 2006a, 2006b

The experience of the law

Orestano 1987, especially pp. 353ff.

1 Medieval Roots

Medieval law

Almeida Costa 1992
Bader and Dilcher 1999
Caravale 1994
Carbasse 2005
Cordes and Kannowski, eds. 2002
Cortese 1995, 2001
Gomes da Silva 2000
Grossi 1995
Guillot and Sassier 2004
Kroeschell and Cordes, eds. 1996

Lupoi 1994
Padoa Schioppa 1995
Schneidmüller and Weinfurter, eds. 2006
Tomas y Valiente 1983

Customary law

Dilcher, ed. 1992
Société Jean Bodin 1990

The early medieval notary

Costamagna, 'L' alto medioevo', in Amelotti and Costamagna 1995
Hilaire 2000, 2003

Canon law

Erdö 2002
Fantappiè 1999
Gaudemet 1994, 1998

The late medieval legal historical context

Dilcher 1996
Grossi 1995
Grossi, 'Il sistema giuridico medievale e la civiltà comunale', in Zorzi, ed. 2008
Quaglioni, 'Introduzione: La rinnovazione del diritto', in Constable and Cracco 2003
Santarelli, 'La normativa statutaria nel quadro dell'esperienza giuridica bassomedievale', in Società italiana di storia del diritto 2001
Zecchino 2005

France

Carbasse 2005
Gouron and Rigaudière, eds. 1988
Rigaudière 2004

Portugal

Hespanha 1999a, 1999b

Spain

Tomas y Valiente 1983

Germany

Kroeschell, Cordes and Nehlsen Von Stryk 2006

Ius commune

Ascheri, Baumgärtner and Kirshner, eds. 1999

Bellomo 1994, 1997, especially the essays 'I giuristi, la giustizia e il sistema del diritto comune'; 'Parlando di diritto comune'; and '*Ius commune*' in the latter
Caravale 2005

Canon law

Ascheri 2000
Erdö 2002
Fantappiè 1999
Gaudemet 1994
Helmholz 1996

Feudalism

Bloch 1939, 1949
Cortese 1995, 2001
Guillot and Sassier 2004

Commercial law

Capitani 1987
Galgano 2001
Hilaire 1986
Petit, ed. 1997
Piergiovanni, ed. 2005
Santarelli 1998

2 The Foundations of the Modern Legal System

Medieval social history

Montanari 2003
Naphy and Spicer 2004, 2006

Medieval intellectual history and the law

Grossi, 'Dalla società di società alla insularità dello stato, tra medioevo ed età moderna', in 2006d
Villey 2003, 2007

Late medieval and early modern France

Carbasse 2005
Krynen 1993
Piano Mortari 1990
Rigaudière 2004

The origins of English common law

Baker 2002; Baker, ed. 2003: II–VI

Caenegem 1988
Caravale 2005
Cavanna 1979
Mattei 2004
Matteucci 2001
Varano, Barsotti 2002

Natural law

Birocchi 2002
Buckle 1991
Costa 1999
Dunn and Harris, eds. 1997
Locke, ed. Laslett 1960
Piovani 1961
Royal Netherland Academy of Arts and Sciences 1984
Van den Bergh 2002
Viano 1997

The legal Enlightenment

Costa 1999
Grossi 2003a, 2004, 2006c, 2007

The bills of rights

Barbera, ed. 1997
Battaglia 1998
Birocchi 2002
Costa 1999
Fioravanti 1999
Rials 1988

The ordonnances

Birocchi 2002
Carbasse 2005
France 1996: I and II
Hilaire 1986
Wieacker 1967, 1980

Constitutional change in seventeenth-century England

Baker, ed. 2003: VI–VIII
Pocock 1987

The French Revolution

Duso 2003
Hoffmann 2003

Hoffmann 2007
Jaume 1990, 2003
Rosanvallon 1998, 2005

Codes and codification

Accademia nazionale dei Lincei 2006
Association Henri Capitant 2004
Cappellini and Sordi 2002
Caroni 1998, 2003
Caroni and Dezza, eds. 2006
Dölmeyer, ed. 1995, 2006
Halpérin 1992, 2001, 2003
Hattenauer, ed. 1970
Martinek, ed. 1999
Petronio 2002
Schubert, ed. 2005
Tomás y Valiente 1983

Early modern German law

Boockmann, ed. 1998–2001: I and II
Diestelkamp 1999
Luig 1998
Oestmann 2002
Schröder 1999, 2001
Wieacker 1967

Pandectism

Cappellini 1984–5
Falk 1989
Haferkamp 2004
Jouanjan 2005
Wieacker 1967, 1974, 1983

The beginnings of public law

Costa and Zolo, eds. 2002
Fioravanti 1979
Mannori and Sordi 2001
Stolleis 1992

Legal solidarity

Halpérin 1992
Ritter 1989, 1996
Tomás y Valiente 1983

Comparative law

Kohler 1885
Maine 1871, 2002
Zweiger and Kötz 1984, 1998a, 1998b

Labour law

Aubin and Bouveresse 1995
Le Goff 2004
Passaniti 2006

Commercial law

Bergfeld 1986
Galgano 2001
Hilaire 1986
Padoa Schioppa 1992
Tomás y Valiente 1983

The codification of canon law

Erdö 2002
Fantappiè 1999
Gaudemet 1994

3 Journeys in Contemporary Law

Corporatism and the Weimar Republic

Costa 2001a, 2001b
Halpérin 2004
Huber 1981, 1984
Kervégan, ed. 2002
Mortati 1946
Stolleis 1999

Russia

Ajani 1996
Cerroni, ed. 1964
Costa 2001b

Fascist Italy and Nazi Germany

Caenegem 2002, 2003
Dreier, ed. 1989
Grossi 2000
Mazzacane, ed. 2005
Schmoeckel 2005
Stolleis 1999, 2006; Stolleis, ed. 1989

Notes

Preliminaries

1 The term *ordinamento* ('order') is central to the work of the great Italian jurist Santi Romano, about whom we shall hear more at the beginning of chapter 3. Romano's concentration on *ordinamento* implies a rediscovery of the *sociality* of the law, of its genesis in the lower echelons of society rather than amongst the high citadels of political power.

1 Medieval Roots

1 For an example of the 'statist' vision of medieval legal history from which I dissent in this section, see Iglesia Ferreiros 1992.
2 John of Salisbury, *Policraticus*, VI. 2.
3 John of Salisbury, *Policraticus*, IV. 2. Aquinas, *Summa theologiae*, II. 2, *quaest.* 58, *art.* 1, *ad* 5.
4 Quoted in Carbasse 2005: 192.
5 Le Goff 2006. The original French edition was entitled *L'Europe est-elle née au moyen age?* ('Was Europe Born in the Middle Ages?') (Le Goff 2003).
6 One of the defining features of the late Middle Ages is the proliferation from the end of the eleventh century onwards of universities as centres for the production and dissemination of research, and for the cultural development of young people.
7 See the title of Calasso's inaugural lecture to the University of Rome, 'Il diritto comune come fatto spirituale', reprinted in Calasso 1951.

8 The text in question is in Bartolo's commentary on the *Digest* and runs as follows: 'quidam Doctor de Aurelianis, ut retulit ille Doctor Theutonicus qui pridie hic repetiit, dicit [...]' ('A German scholar reports that a certain Doctor of Orléans, whom he quoted earlier today, has said ...') (fragment 17, heading 2, on book 41).

9 For an example of Calasso's work, see especially Calasso 1954.

2 The Foundations of the Modern Legal System

1 'Velut in confinio duorum populorum constitutus, ac simul ante retroque prospiciens' ('as if I were standing at the border between two peoples, and were looking both forwards and backwards') (Petrarch, *Rerum memoranordum libri*, I. 19).

2 A phrase used by St Louis's chancellery and quoted in Rigaudière 2004: 93.

3 The *Songe du Vergier* (Schnerb-Lièvre, ed. 1982) is a work written between 1376 and 1378, by order of King Charles V, first in Latin and then the French vernacular, which bears faithful witness to the outlook of the fourteenth-century French monarchy. The quotation, 'faire loys ou constitucions toutes nouvelles entre ses subjés', can be found in Rigaurdière 2004: 125.

4 The distinction can be found in *Les six livres de la République*, I. 8.

5 The *Summa angelica de casibus conscientiae* ('Angelic Summary of Cases of Conscience'), was first published in Venice in 1487. It was a very well-read text, reprinted many tens of times over the next few centuries, even as late as 1771, in Rome.

6 Alberti was also the author of a short tractate entitled *De iure* ('On Law'), further demonstrating his pansophism. For Alberti and for legal humanism in general, see Quaglioni 2004, especially ch. 9, 'L'umanesimo e la giustizia' ('Humanism and Justice').

7 The thinkers most associated with this 'mathematical' view of the law are the German philosopher-jurists Gottfried Wilhelm Leibniz (1646–1716) and Christian Wolff (1670–1754).

8 If the nature of this book did not demand sacrifices in favour of a more coherent account of the topic, I would have liked to have included here an account of Second Scholasticism – a philosophical current made up of theologian-jurists operating mainly in Spain in the sixteenth and seventeenth centuries. These thinkers made a great contribution to shaping the law along the lines of the new post-humanist anthropology. Indeed, Grotius and the whole natural law movement are, to an extent, mere disciples of Second Scholasticism. See Grossi, ed. 1973.

9 For van Groot's *De iure belli ac pacis*, see the edition by Feenstra and Persenaire: van Groot 1993). The original Latin of the two texts quoted above runs as follows: 'Primum mihi cura haec fuit, ut eorum quae ad ius naturae pertinent probationes referrem ad notiones quasdam tam certas ut eas nemo negare possit, nisi sibi vim inferat. Prinicipia enim eius iuris [...] per se patent atque evidentia sunt' (*Prolegomena*, XXXIX); 'Vere enim profiteor, sicut mathematici figuras a corporibus semotas considerant, ita me in iure tractando ab omni singulari facto abduxisse animum' (*Prolegomena*, LVIII).

10 The decade 1680–90 includes such momentous events in English history as the Glorious Revolution of 1688, in which the absolutist monarch James II was forced to flee the country; the Bill of Rights of 1689, the first document of its kind, which bound the new monarch to respect certain rights of his subjects; and the publication in 1690 of Locke's above-mentioned *Two Treatises*.

11 For more on the *Lei da Boa Razão*, see Almeida Costa 1992.

12 The first economists were arguably the 'physiocrats', a current of thought arising in France in the mid seventeenth century ('physiocrat' is a composite Greek word implying 'government by nature'). The followers of the movement rejected the conflation of the economic sphere with the political and the view of economics as a simple tool in the hands of those who wielded power. Instead they sought to set up a new discipline founded in what they saw as the authentic nature of society. The physiocrats were thus a 'new entrant on the European intellectual stage', because they were the first to research the principles and autonomous rules of the economic aspects of society.

13 For more on this historical episode of legal absolutism, see Grossi 1998.

14 'Legal despotism' is a term deployed primarily by the physiocrats themselves. In their view the coincidence between the interpretation of nature and political power is necessary in order that laws 'need be nothing more than the natural results of order sealed with the stamp of public authority', as Mercier de la Rivière puts it in an essay with the emblematic title 'L'Ordre naturel et essentiel des sociétés politiques' ('The Natural and Essential Order of Political Societies'). The original quotation runs 'ne doivent être que des résultats évidents de l'ordre mais scellés du sceau de l'autorité publique' (see Capitani's essay 'I fisiocrati e Mably tra dispotismo legale e governo misto', in Felice, ed. 2001).

15 Muratori, *Dei difetti della giurisprudenza*, XX.

16 Ibid.

17 Beccaria, *Dei delitti e delle pene*, IV.

18 A declaration made by Pietro Verri, one of the most active exponents of the Enlightenment in Lombardy, in his *Memorie storiche sulla economia pubblica dello Stato di Milano* ('Historical Notes on the Public Economy of the State of Milan'); see Verri 1804: 170.

19 These ideas are found throughout Rousseau's works. See, for example, *Du contrat social ou principes du droit politique* (1762) in Rousseau 1971: II.

20 See the essay 'Beantwortung der Frage: Was ist Aufklärung?', in Sasso, ed. 1977. See also paragraph XLIX of the section 'Rechtslehre' ('Doctrine of Right'), in the *Metaphysik der Sitten* ('Metaphysics of Morals'), in Kant 1968, 1996, 1999.

21 A *recopilación* collects and orders a series of scattered laws and is thus a work of consolidation, rather than legislation proper. From the end of the sixteenth century onwards, *recopilaciones* were carried out in the various Spanish kingdoms on the initiative either of officials or of a local jurist. Their story begins with the Castilian *Nueva recopilación*, ordered by Philip II in 1567, and ends with the *Novísima recopilcacíon*, ordered by Charles IV in 1805. The latter is a massive collection of legislation, divided into twelve volumes, which was widely criticized since it was released at a historical juncture in which all of Europe was turning towards the radical innovation of legal codes as a way of systematizing sources of law. The Spanish efforts at consolidation appeared backwards, given that the year before Napoleon had promulgated the *Code civil*, an extremely novel piece of legislation.

22 The Le Chapelier Declaration (4–17 June 1791) states that: 'L'anéantissement de toutes les espèces de corporations des citoyens du même état et profession, étant une des bases fondamentales de la constitution française, il est défendu de les rétablir de fait, sous quelque prétexte et quelque forme que ce soit' ('Given that the abolition of corporations of citizens of the same status and profession is one of the fundaments of the French Constitution, citizens shall be forbidden from re-establishing such bodies in fact, under whatever pretext and in whatever guise').

23 Necker's text can be found in Necker 1820–1: 435.

24 The historian in question is Mario Viora, who wrote an essay putting forward this point of view in 1928 that has been extraordinarily influential, at least in Italy. See the reprinted edition, Viora 1990.

25 De Tocqueville, *L'Ancien Régime et la révolution*, I. 2, ch. 1, note entitled 'Code du Grand Frédéric'. See De Tocqueville 1955, 1989.

26 Translated from the official Italian-language version of the ABGB for the Viceroyalty of Lombardy-Venetia.

27 Article 2 of the Italian Civil Code of 1865 says the following: 'Communes [divisions of local government], provinces, civil, public or religious bodies and all legally recognized moral bodies in general shall be considered as legal persons and shall be accorded such civil rights as are established by legislation and the practice of public law'.
28 The title of Pellegrino Rossi's famous symposium is 'Observations sur le droit civil français dans ses rapports avec l'état économique de la société' ('Observations on the Relations between French Civil Law and the Economic Situation of Society'), in Rossi 1867: II.
29 *Über di Notwendigkeit eines allgemeinen bürgerlichen Rechts in Deutschland* (Heidelberg, 1814). The original German-language text can be found reprinted in Hattenauer, ed. 2002.
30 Savigny and Thibaut 1982: 79.
31 *Vom Beruf unsrer Zeit für Gesetzgebung und Rechtswissenschaft* (Heidleberg, 1814). See Hattenhauer, ed. 2002; Savigny and Thibaut 1982.
32 Savigny and Thibaut 1982: 97.
33 The title of Savigny's later systematizing work is *System des heutigen römischen Rechts* (Berlin, 1840–6). See Savigny 1886–98, 1981. For the quotations which follow in the text, see Savigny 1886–98: I, pp. 5, 185, 49, 63.
34 Gerber's *Grundzüge eines Systems des deutschen Staatrechts* ('Outline of a System of German Constitutional Law'; 1st published Leipzig, 1865; 2nd edn., Leipzig, 1869; 3rd edn., Dresden, 1880) talks of a 'period of transition' (Gerber 1971: 103).
35 The journal, founded by Gerber and Jhering in 1857, is called *Jahrbucher für die Dogmatik des heutigen römischen und deutschen Privatrechts*.
36 The letter can be found in Losano, ed. 1984: 18.
37 Gerber 1971: 95.
38 Gerber 1971: 97.
39 Gerber 1971: 124.
40 The reference to Otto von Gierke is to the *Entwurf eines Bürgerlichen Gestzbuchs und das deutsche Recht* ('The Draft Civil Code and German Law', 1888–9), and also to *Die soziale Aufgabe des Privatrechts* ('The Social Function of Private Law', 1889).
41 The work in question by Menger is *Das Bürgerliche Recht und die besitzlosen Volksklassen: Eine Kritik des Entwurfs eines Bürgerlichen Gesetzbuchs für das Deutsche Reich* ('Civil law and the Proletariat: A Critique of the Draft Civil Code for the Kingdom of Germany'); see Menger 1890.
42 See Cruet 1908.

43 There was even a doctoral thesis written on the 'Praetorian Power of Jurisprudence', see Langlois 1897.

44 The quotations from Saleilles here can be found in Grossi 1998: 219, 217. For the 'frame', see ibid. 239. On the Napoleonic Code as a legal system in need of massive interventions from judges, see ibid. 251, especially n. 220. Saleilles's emblematic text is 'Le Code Civil et la méthode historique' ('The Civil Code and the Historical Method'), which can be found in Halpérin 2004.

45 There is an entire journal issue dedicated to the consideration of Gény's work: *Quaderni fiorentini per la storia del pensiero giuridico moderno*, 20 (1991).

46 See Gény 1899. (There is also a second edition from 1919, which leaves the original text unchanged but incorporates a number of notable additions, including one on the free-law doctrine, which we will examine in the next section.)

47 See Paolo Grossi, 'Ripensare Gény', in Grossi 1998: 158, 161 and 172.

48 All these authors' works have now been reprinted: see Fuchs 1970–5; Ehrlich 1973; Kantorowicz 1988, 2002.

49 Kantorowicz 1988: 80–1, 82.

50 Gabba 1901: 50.

51 The liberal journalist Heinrich Bernhard Oppenheim was the first to coin this label in an article in the 7 December 1871 edition of the *Nazionalzeitung*. *Kathedersozialismus*, literally 'socialism of the university chair', indicates a programme of left-leaning politics advanced by members of the academe.

52 The article in question can be found in Marx and Engels 1964–8: XXI, 491ff.

53 For more information on this current in legal thought, see two numbers of the *Quaderni fiorentini per la storia del pensiero giuridico moderno*, 3 (1974), and 4 (1975).

54 For Menger's essay see n. 41 above.

55 Treves 1894.

56 In 1903, Emmanuel Levy published a book in Paris with just such a title: *L'Affirmation du droit collectif*.

57 Saleilles 1888–9a, 1888–9b, 1901, 1904.

58 Saleilles 1900.

59 Saleilles 1910.

60 See Rabel 1965.

61 See Lambert 1903, 1919, 1921.

62 The text here is a translation of the official Italian version of the Napoleonic Code (1806).

63 On Barassi and Italian labour law, see Napoli, ed. 2003.
64 On Lotmar and Hugo Sinzmeyer, whom we shall encounter below, see Becker 1995.
65 See Lotmar 1902. The reference to *faktische Umwelt* is on p. 24 of the introduction. For the labour contract as a personal, not a proprietary matter, see p. 8.
66 The essay in question, 'Die Tarifverträge zwischen Arbeitgebern und Arbeitnehmern' ('Tariff Contracts between Employers and Employees'), together with several other contributions on the topic, can be found reprinted in Lotmar 1992.
67 Sinzheimer's main works can be found in the collection Kahn-Freund, ed. 1976.
68 The fundamental piece of scholarship on bills of exchange is Einert 1969, originally published in Leipzig in 1839. Carl Einert's work systematizes a practice that was already common in everyday business dealings.
69 Goldschmidt 1913: 12–13.
70 Cited in Tomas y Valiente 1983: 518.

3 Journeys in Contemporary Law

1 The speech can be found in Santi Romano 1969. The reference to the 'eclipse' of the state is on p. 9, whilst the accusation of oversimplification is on p. 13.
2 See Treves 1894.
3 See Romano 1977. This is the 3rd edition of the work, which was first published in Pisa in 1918. The 2nd edition, printed in Florence in 1946, should also be mentioned because it was edited by the author himself who responded therein to many of the criticisms levelled at him in the intervening years. See also the Spanish, French and German translations: Romano 1963, 1975a, 1975b.
4 Published in Leipzig in 1843.
5 Gierke 1954; the first volume was published in Berlin in 1868. There is a French translation of the first volume available by Jean de Pange (Gierke 1914), and an English one by Frederic Maitland (Gierke 1900).
6 See Preuss 1964; the volume was originally published in Berlin in 1889, under the title *Gemeinde, Staat, Reich als Gebietskörperschaften: Versuch einer deutschen Staatskonstruktion auf Grundlage der Genossenschaftstheorie* ('Community, State and Empire as Territorial Associations: An Attempt to Construct a German State Based on the Theory of Associations').

7 Albert Schäffle's fixation with biology is apparent even from the title of his great work, which translates as 'The Structure and Life of the Social Body' (see Schäffle 1875–8). There is also an Italian translation: Schäffle 1881.

8 Bishop Ketteler published an essay in 1864 entitled *Die Arbeitfrage und das Christentum* ('The Question of the Workers and Christianity').

9 Toniolo's most representative writings are: 'Programma dei cattolici di fronte al socialismo' ('The Catholic Response to Socialism', 1894); 'Indirizzi e concetti sociali all'esordire del secolo XX' ('Social Trends and Concepts at the Beginning of the Twentieth Century', 1900); 'La democrazia cristiana' ('Christian Democracy', 1900); and 'Provvedimenti sociali popolari' ('Social Provisions for the People', 1902). These can be found in Toniolo 1951.

10 On the birth of the notion of business enterprises in early twentieth-century Europe, see Grossi 2000: 191ff. This is the time at which we can first talk of 'economic law', reflecting a new-found concentration by jurists on economic realities. For example, in the aftermath of the First World War, one German jurist who was sensitive to socio-economic change, Justus-Wilhelm Hedemann, set up an Institut für Wirtschaftsrecht (Institute for Economic Law) at the University of Jena. See also his book setting out an intellectual programme for the new discipline: Hedemann 1922.

11 The quotations are taken from an essay entitled 'Diritto di guerra e diritto di pace' ('The Law of War and the Law of Peace', 1918), now in Ferrara 1954: I, pp. 70–1.

12 Gierke lived until 1921.

13 The 'socialized person', or *Verbandmensch*, is a term coined by Friedrich Naumann, one of the leaders of the Weimar project, in an extremely lucid and persuasive contribution to the Constitutional Assembly. See the proceedings of the assembly: German National Assembly 1920: 179–80.

14 See Gierke 1973: 31. The book was originally published in Aalen in 1915.

15 The expression *collective democracy* comes from Ernst Fraenkel (1898–1975), a pupil of Sinzheimer at Frankfurt who became a lawyer and a trade unionist. The essay, 'Kollektive Demokratie' (1929), can be found in Fraenkel 1966.

16 I should remember here the famous and oft-quoted line on which Article 153 of the Constitution, dealing with property, ends: 'Eigentum verpflichtet. Sein Gebrauch soll zugleich Dienst sein für das Gemeine Beste' (roughly translated: 'Property brings duties. Its use should also serve the common good.')

17 'Self-regulation' here translates the German *Selbstgesetzgebung*, which is Sinzheimer's way of defending the position of the trade union within the state and the community.
18 Quoted in Ajani 1996, which also gives a much broader account of the Russian legal context.
19 See Lenin 1974.
20 East Germany would also draft a constitution in 1949, setting out its own version of a socialist system for state and society.

Bibliography

Accademia nazionale dei Lincei, *Il bicentenario del Codice napoleonico* (Rome: Bardi, 2006).

Ajani, Gianmaria, *Diritto dell'Europa orientale* (Torino: UTET, 1996).

Almeida Costa, Mário Júlio de, *Historia do direito portugês* (Coimbra: Almedina, 1992).

Amelotti, Mario, and Giorgio Costamagna, *Alle origini del notariato italiano* (Rome: Consiglio nazionale del notariato: 1975; repr. Milan: Giuffrè, 1995).

Ancel, Marc, ed., *Livre du centenaire de la Société de legislation comparée: Un siècle de droit comparé en France, 1869–1969* (Paris: Librairie générale de droit et de jurisprudence, 1969).

Ascheri, Mario, *I diritti del Medioevo italiano: Secoli XI–XV* (Rome: Carocci, 2000).

Ascheri, Mario, Ingrid Baumgärtner and Julius Kirshner, eds., *Legal Consulting in the Civil Law Tradition* (Berkelely: Robbins Collection, 1999).

Aubin, Gérard, and Jacques Bouveresse, *Introduction historique au droit du travail* (Paris: Presses universitaires de France, 1995).

Bader, Karl, and Gerhard Dilcher, *Deutsche Rechtsgeschichte: Land und Stadt-Bürger und Bauer im Alten Europa* (Berlin: Springer, 1999).

Baker, J. H., *An Introduction to English Legal History*, 4th edn. (London: Butterworth, 2002).

Baker, J. H., ed., *The Oxford History of the Laws of England*, 6 vols. (Oxford: Oxford University Press, 2003).

Barbera, Augusto, ed., *Le basi filosofiche del costituzionalismo: Lineamenti di filosofia del diritto costituzionale* (Rome and Bari: Laterza, 1997).

Battaglia, Felice, Augusto Barbera and Nicola Matteucci, with Francesca Rescigno, *Le carte dei diritti*, 3rd edn., ed. Alberto Calogero and Claudio Carbone (Reggio Calabria: Laruffa, 1998).

Becker, Martin, *Arbeitsvertrag und Arbeitsverhältnis in Deutschland: Vom Beginn der Industrialisierung bis zum Ende des Kaiserreichs* (Frankfurt: Klostermann, 1995).

Bellomo, Manlio, *L'Europa del diritto comune*, 7th edn. (Rome: Il cigno Galileo Galilei, 1994).

Bellomo, Manlio, *Medioevo edito e inedito*, 3 vols. (Rome: Il cigno Galileo Galilei, 1997), II: *Scienza del diritto e società medievale.*

Bergfeld, Christoph, 'Handelsrecht', in *Handbuch der Quellen und Literatur der Neueren Europäishe Privatrechtsgeschichte*, ed. Helmut Coing, 3 vols. (Munich: Beck, 1976–88), III/3: *Gesetzbung zu den privatrechtlichen Sondergebieten* (1986).

Bergh, Govaert C. J. J. van den, *Die holländische elegante Schule: Ein Betrag zur Geschichte von Humanismus und Rechtswissenschaft in den Niederlanden 1500–1800* (Frankfurt: Klostermann, 2002).

Birocchi, Italo, *Alla ricerca dell'ordine: Fonti e cultura giuridica nell'età moderna* (Turin: Giappichelli, 2002).

Bloch, Marc, *La Société féodale* (Paris: Michel, 1939).

Bloch, Marc, *La società feudale*, trans. Bianca Maria Cremonesi (Turin: Einaudi, 1949).

Bodin, Jean, *Les Six Livres de la République* (Aalen: Scientia, 1977).

Boockmann, Hartmut, ed., *Neuzeit: Recht und Verfassung im Übergang vom Mittelalter zur Neuzeit*, 2 vols. (Göttingen: Vandenhoek & Ruprecht, 1998–2001).

Braudel, Fernand, *L'Europe: L'Espace, le temps, les hommes* (Paris: Arts et métiers graphiques, 1987).

Buckle, Stephen, *Natural Law and the Theory of Property: Grotius to Hume* (Oxford: Clarendon Press, 1991).

Caenegem, Raoul C. van, *The Birth of the English Common Law*, 2nd edn. (Cambridge: Cambridge University Press, 1988).

Caenegem, Raoul C. van, *European Law in the Past and the Future: Unity and Diversity over Two Millennia* (Cambridge: Cambridge University Press, 2002).

Caenegem, Raoul C. van, *I sistemi giuridici europei*, trans. Emmanuela Bertucci (Bologna: Il Mulino 2003).

Calasso, Francesco, *Introduzione al diritto comune* (Milan: Giuffrè, 1951).

Calasso, Francesco, *Medio evo del diritto* (Milan: Giuffrè, 1954).

Capitani, Ovidio, *Una economia politica nel Medioevo* (Bologna: Pàtron, 1987).

Cappellini, Paolo, *Systema iuris*, 2 vols. (Milan: Giuffrè, 1984–5).

Cappellini, Paolo, and Bernardo Sordi, *Codici: Una riflessione a fine di millennio* (Milan: Giuffrè, 2002).

Caravale, Mario, *Ordinamenti giuridici dell'Europa medievale* (Bologna: Il Mulino, 1994).

Caravale, Mario, *Alle origini del diritto europeo: Ius commune, droit commun, common law, nella dottrina giuridica della prima età moderna* (Bologna: Monduzzi, 2005).

Carbasse, J. M., *Manuel d'introduction historique au droit* (Paris: Presses universitaires de France, 2005).

Caroni, Pio, *Saggi sulla storia della codificazione* (Milan: Giuffrè, 1998).
Caroni, Pio, *Gesetz und Gesetzbuch: Beiträge zu einer Kodifikationsgeschichte*, trans. Karin Caroni (Basel: Helbing & Lichtenhahn, 2003).
Caroni, Pio, and Ettore Dezza, *L'ABGB e la codificazione asburgica in Italia e in Europa* (Padova: CEDAM, 2006).
Cavanna, Adriano, *Storia del diritto moderno in Europa: Le fonti e il pensiero giuridico* (Milan: Giuffrè, 1979).
Cerroni, Umberto, ed., *Teorie sovietiche del diritto* (Milan: Giuffrè, 1964).
Chabod, Federico, *Storia dell'idea d'Europa* (Bari: Laterza, 1964).
Cordes, Albrecht, and Bernt Kannowski, eds., *Rechtsbegriffe im Mittelalter* (Frankfurt: Peter Lang, 2002).
Cortese, Ennio, *Il diritto nella storia medievale*, 2 vols. (Rome: Il cigno Galileo Galilei, 1995), I: *L'alto Medioevo*; II: *Il basso Medioevo*.
Cortese, Ennio, *Le grandi linee della storia giuridica medievale*, (Rome: Il cigno Galileo Galilei, 2001).
Costa, Pietro, *Civitas: Storia della cittadinanza in Europa*, 4 vols. (Rome and Bari: Laterza, 1999–2001), I: *Dalla civiltà comunale al Settecento* (1999).
Costa, Pietro, *Civitas: Storia della cittadinanza in Europa*, 4 vols. (Rome and Bari: Laterza, 1999–2001), III: *La civiltà liberale* (2001a).
Costa, Pietro, *Civitas: Storia della cittadinanza in Europa*, 4 vols. (Rome and Bari: Laterza, 1999–2001), IV: *L'età dei totalitarismi e della democrazia* (2001b).
Costa, Pietro, and Danilo Zolo, eds., *Lo Stato di diritto: Storia, teoria, critica* (Milan: Feltrinelli, 2002).
Constable, Giles, and Giorgio Cracco, eds., *Il secolo XIII: La 'renovatio' dell'Europa cristiana* (Bologna: Il Mulino, 2003).
Cour de Cassation, Ordre des avocats au Conseil d'État et à la Cour de cassation and Association Henri Capitant des amis de la culture juridique française, *Le Code Civil 1804–2004: Livre du bicentenaire* (Paris: Dalloz; LexisNexis Litec, 2004).
Cour de Cassation, Direction des archives, Direction des journaux officielles, *Les Français et leur Code Civil* (Paris: Éditions des journaux officielles, 2004).
Cruet, Jean, *La Vie du droit et l'impuissance des lois* (Paris: Flammarion, 1908).
De Tocqueville, Alexis, *The Old Regime and the Revolution*, trans. Stuart Gilbert (New York: Anchor, 1955).
De Tocqueville, Alexis, *L'antico regime e la rivoluzione*, ed. Giorgio Candeloro (Milan: Rizzoli, 1989).
Diestelkamp, Bernhard, *Recht und Gericht im Heiligen Römischen Reich* (Frankfurt: Klostermann, 1999).
Dilcher, Gerhard, ed., *Gewohnheitsrecht und Rechtsgewohnheiten im Mittelalter* (Berlin: Duncker & Humblot, 1992).
Dilcher, Gerhard, *Bürgerrecht und Stadtverfassung im europäischen Mittelalter* (Cologne: Böhlau, 1996).
Dölmeyer, Barbara, ed., *200 Jahre Landrechts für die Preussischen Staaten: Wirkungsgeschichte und internationaler Kontext* (Frankfurt: Klostermann, 1995).

Dölmeyer, Barbara, ed., *Richterliche Anwendung des Code Civil in seinen europäischen Geltungsbereichen auserhalb Frankreichs* (Frankfurt: Klostermann, 2006).

Dreier, Ralf, ed., *Recht und Justiz im 'Dritten Reich'* (Frankfurt: Suhrkamp, 1989).

Dunn, John, and Ian Harris, eds., *Grotius* (Cheltenham: Elgar, 1997).

Duso, Giuseppe, *La rappresentanza politica: Genesi e crisi del concetto*, 2nd edn. (Milan: Giuffrè, 2003).

Ehrlich, Eugen, *Frie Rechtsfindung und freie Rechtswissenschaft* (Aalen: Scientia, 1973).

Einert, Carl, *Das Wechselrecht nach dem Bedürfnis des Wechselgeschäfts* (Aalen: Scientia, 1969).

Erdö, Peter, *Die Quellen des Kirchenrechts: Eine geschichtliche Einführung* (Frankfurt: Peter Lang, 2002).

Falk, Ulrich, *Ein Gelehrter wie Windscheid: Erkundungen auf den Feldern der sogenannten Begriffsjurisprudenz* (Frankfurt: Klostermann, 1989).

Falk, Ulrich, ed., *Das Bürgerliche Gesetzbuch und seine Richter: Zur Reaktion der Rechtssprechung auf die Kodifikation des deutschen Privatrechts 1896–1914* (Frankfurt: Klostermann, 2002).

Fantappiè, Carlo, *Introduzione storica al diritto canonico* (Bologna: Il Mulino, 1999).

Felice, Domenico, ed., *Dispotismo: Genesi e sviluppi di un concetto filosofico-politico* (Naples: Liguori, 2001).

Ferrara, Francesco, *Scritti giuridici*, 3 vols. (Milan: Giuffrè, 1954).

Fioravanti, Maurizio, *Giuristi e costituzione politica nell'Ottocento tedesco* (Milan: Giuffrè, 1979).

Fioravanti, Maurizio, *Costituzione* (Bologna: Il Mulino, 1999).

Fraenkel, Ernst, *Arbeitsrecht und Politik: Quellentexte 1918–1933*, ed. Thilo Ramm (Berlin and Neuwied: Luchterhand, 1966).

France, Kingdom of, *Code Louis*, 2 vols. (Milan: Giuffrè, 1996), I: *Ordannance Civile* (1667).

France, Kingdom of, *Code Louis*, 2 vols. (Milan: Giuffrè, 1996), II: *Ordannance Criminelle* (1670).

Fuchs, Ernst, *Gesammelte Schriften über Freirecht und Rechtsreform*, ed. Albert S. Foulkes, 3 vols. (Aalen: Scientia, 1970–5).

Gabba, Bassano, *Trenta anni di legislazione sociale* (Turin: Bocca, 1901).

Galgano, Francesco, *Lex mercatoria: Storia di diritto commerciale* (Bologna: Il Mulino, 2001).

Gaudemet, Jean, *Église et cité: Histoire du droit canonique* (Paris: Cerf/Montchrestien, 1994).

Gaudemet, Jean, *Storia del diritto canonico: Ecclesia et Civitas*, trans. Alessandra Ruzzon and Tiziano Vanzetto (Cinisello Balsamo: San Paolo, 1998).

Gény, François, *Méthode d'interprétation et sources en droit privé positif* (Paris: Bibliothèque de jurisprudence civile contemporaine, 1899).

Gerber, Carl Friedrich von, *Diritto pubblico*, ed. Pier Luigi Lucchini (Milan: Giuffrè, 1971).

German National Assembly, *Verhandlungen der Verfassunggebenden deutschen Nationalversammlung* (Berlin: Norddeutsche Buchdrucherei, 1920).
Gierke, Otto von, *Political Theories of the Middle Age*, trans. Frederic William Maitland (Cambridge: Cambridge University Press, 1900).
Gierke, Otto von, *Théories politiques du Moyen Âge*, trans. Jean de Pange (Paris: Tenin, 1914).
Gierke, Otto von, *Das Deutsche Genossenschaftsrecht* (Graz: ADEVA, 1954).
Gierke, Otto von, *Die Grundbegriffe des Staatsrechts und die neuesten Staatsrechtstheorien* (Aalen: Scientia, 1973).
Goldschmidt, Leon, *Universalgeschichte des Handelsrechts* (Stuttgart: Enke, 1891).
Goldschmidt, Leon, *Storia universale del diritto commerciale*, trans. Vittorio Pouchain and Antonio Scialoja (Turin: UTET, 1913).
Gomes da Silva, Nuno Espinosa, *Historia do direito portugês: Fontes de direito* (Lisbon: Fundação Calouste Gulbenkian, 2000).
Gouron, André, and Albert Rigaudière, eds., *Renaissance du pouvoir législatif et genèse de l'État* (Montpellier: Socapress, 1988).
Groot, Hugh van, *De iure belli ac pacis libri tres*, ed. Robert Feenstra and C. E. Persenaire (Aalen: Scientia, 1993).
Grossi, Paolo, ed., *La Seconda Scolastica nella formazione del diritto privato moderno* (Milan: Giuffrè, 1973).
Grossi, Paolo, *L'ordine giuridico medievale* (Rome; Bari: Laterza, 1995).
Grossi, Paolo, *Assolutismo giuridico e diritto privato* (Milan: Giuffrè, 1998).
Grossi, Paolo, *Scienza giuridica italiana: Un profilo storico 1860–1950* (Milan: Giuffrè, 2000).
Grossi, Paolo, *Mitología jurídica de la modernidad*, trans. Manuel Martínez Neira (Madrid: Trotta, 2003a).
Grossi, Paolo, *Prima lezione di diritto* (Rome; Bari: Laterza, 2003b).
Grossi, Paolo, *Mitologias jurídicas da modernidade*, trans. Arno Dal Ri Júnior (Florianopolis: Fundação Boiteaux, 2004).
Grossi, Paolo, *Primeira lição sobre direito*, trans. Ricardo Marselo Fonseca (Rio de Janeiro: Forense, 2006a).
Grossi, Paolo, *La primera lección de derecho*, trans. Clara Álvarez Alonso (Madrid: Marcial Pons, 2006b).
Grossi, Paolo, *La proprietà e le proprietà nell'officina dello storico* (Naples: Editoriale scientifica, 2006c).
Grossi, Paolo, *Società, diritto, stato: Un recupero per il diritto* (Milan: Giuffrè, 2006d).
Grossi, Paolo, *Mitologie giuridiche della modernità*, 3rd edn. (Milan: Giuffrè, 2007).
Guillot, Olivier, and Yves Sassier, *Pouvoirs et institutions dans la France médiévale: Des origines à l'époque féodale* (Paris: Armand Colin, 2004).
Haferkamp, Hans Peter, *Georg Friedrich Puchta und die 'Begriffsjurisprudenz'* (Frankfurt: Klostermann, 2004).
Halpérin, Jean-Louis, *L'Impossible Code civil* (Paris: Presses universitaires de France, 1992).

Halpérin, Jean-Louis, *Histoire du droit privé français depuis 1804*, 2nd edn. (Paris: Presses universitaires de France, 2001).

Halpérin, Jean-Louis, *Le Code civil*, 2nd edn. (Paris: Dalloz, 2003).

Halpérin, Jean-Louis, *Histoire des droits en Europe: De 1750 à nos jours* (Paris: Flammarion, 2004).

Hattenauer, Hans, ed., *Allgemeines Landrecht für di preussischen Staaten von 1794*, 2 vols. (Frankfurt: Metzner, 1970–3), I: *Textausgabe* (1970).

Hattenauer, Hans, ed., *Thibaut und Savigny: Ihre programmatischen Schriften*, 2nd edn. (Munich: Vahlen, 2002).

Hedemann, Justus-Wilhelm, *Grundzüge des Wirtschaftsrecht* (Mannheim: Bensheimer, 1922).

Helmholz, R. H., *The Spirit of Classical Canon Law* (Athens, GA and London: University of Georgia Press, 1996).

Hespanha, Antònio Manuel, *Panorama histórico da cultura jurídica europeia* (Lisbon: Universidade Autónoma Luís de Camõe, 1999a).

Hespanha, Antònio Manuel, *Introduzione alla storia del diritto europeo*, trans. Livia Apa and Leonardo Santi (Bologna: Il Mulino 1999b).

Hilaire, Jean, *Introduction historique au droit commercial* (Paris: Presses universitaires de France, 1986).

Hilaire, Jean *La science des notaires: Une longue histoire* (Paris: Presses universitaires de France, 2000).

Hilaire, Jean, *La scienza dei notai: La lunga storia del notariato in Francia* (Milan: Giuffrè, 2003).

Hoffmann, Hasso, *Repräsentation: Studien zur Wort und Begriffsgeschichte bis ins 19. Jahrhundert*, 4th edn. (Berlin: Duncker & Humblot, 2003).

Hoffmann, Hasso, *Rappresentanza-rappresentazione: Parola e concetto dall'antichità all'Ottocento*, trans. Claudio Tommasi (Milan: Giuffrè, 2007).

Huber, Ernst Rudolf, *Deutsche Verfassungsgeschichte seit 1789*, 8 vols. (Stuttgart: Kohlhammer, 1957–91), V: *Weltkrieg, Revolution und Reichserneuerung* (1978).

Huber, Ernst Rudolf, *Deutsche Verfassungsgeschichte seit 1789*, 8 vols. (Stuttgart: Kohlhammer, 1957–91), VI: *Die Weimarer Reichsverfassung* (1981).

Huber, Ernst Rudolf, *Deutsche Verfassungsgeschichte seit 1789*, 8 vols. (Stuttgart: Kohlhammer, 1957–91), VII: *Ausbau, Schutz und Untergang der Weimarer Republik* (1984).

Iglesia Ferreiros, Aquilino, *La creación del derecho: Una historia de la formacion de un derecho estatal español. Manual*, 2 vols. (Barcelona: Signo, 1992).

Jaume, Lucien, *Échec au liberalisme: Les Jacobins et l'État* (Paris: Kimé, 1990).

Jaume, Lucien, *Scacco al liberalismo: I giacobini e lo stato*, trans. Cinzia Cassani (Naples: Editoriale scientifica, 2003).

Jouanjan, Olivier, *Une histoire de la pensée juridique en Allemagne, 1800–1918: Idéalisme e conceptualisme chez les juristes allemands du 19e siècle* (Paris: Presses universitaires de France, 2005).

Kahn-Freund, Otto, ed., *Arbeitsrecht und Rechtssociologie: Gesammelte Aufsätze und Reden von Hugo Sinzheimer*, 2 vols. (Frankfurt: Klostermann, 1976).

Kant, Immanuel, *Werke*, ed. Wilhelm Weischedel, 12 vols. (Frankfurt: Suhrkamp, 1968).

Kant, Immanuel, *Metaphysics of Morals*, trans. Mary J. McGregor (Cambridge: Cambridge University Press, 1996).

Kant, Immanuel, *Metafisca dei costumi*, trans. Nicolao Merker, 6th edn. (Rome and Bari: Laterza, 1999).

Kantorowicz, Herman, *La lotta per la scienza del diritto*, trans. Raffaele Majetti (Bologna: Forni, 1988).

Kantorowicz, Herman, *Der Kampf um die Rechtswissenschaft* (Baden-Baden: Nomos, 2002).

Kervégan, Jean-François, ed., *Crise e pensée de la crise en droit: Weimar, sa république et ses juristes* (Lyon: ENS, 2002).

Kohler, Joseph, *Einleitung in di vergleichende Rechtswissenschaft* (Würzburg: Stahel, 1885).

Kroeschell, Karl, and Albrecht Cordes, eds., *Funktion und Form: Quellen und Methodenprobleme der mittelalterlichen Rechtsgeschichte* (Berlin: Duncker & Humblot, 1996).

Kroeschell, Karl, Albrecht Cordes and Karen Nelson-von Stryk, *Deutsche Rechtsgeschichte*, 3 vols. (Cologne: Böhlau, 2005–8), II: *1250–1650* (2006).

Krynen, Jacques, *L'Empire du roi: Idées et croyances politiques en France, XIIIe–XVe siècle* (Paris: Gallimard, 1993).

Lambert, Édouard, *Études de droit commun législatif: La fonction du droit civil comparé.* (Paris: Giard & Brière, 1903).

Lambert, Édouard, *L'Enseignement du droit civil comparé: Sa cooperation au rapprochement entre la jurisprudence française et la jurisprudence anglo-americaine* (Lyon: Rey, 1919).

Lambert, Édouard, *L'Institut de droit comparé: Son programme; les méthodes d'enseignement* (Lyon: Rey, 1921).

Langlois, Henri, *Essai sur le pouvoir prétorien de la jurisprudence en droit français* (Caen: Valin, 1897).

Le Goff, Jacques, *L'Europe est-elle née au moyen age?* (Paris: Seuil, 2003).

Le Goff, Jacques, *Du silence à la parole: Une histoire du droit du travail des années 1830 à nos jours* (Rennes: Presses universitaires de Rennes, 2004).

Le Goff, Jacques, *The Birth of Europe: 400–1500* (Oxford: Blackwell, 2006).

Lenin, Vladimir Ilyich, 'Stato e rivoluzione', in *Opere scelte*, 6 vols. (Rome: Riuniti, 1973–5), IV (1974).

Lequette, Yves, and Laurent Leveneur, *Le Code civil, 1804–2004: Un passé, un présent, un avenir* (Paris: Dalloz, 2004).

Locke, John, *Two Treatises of Government*, ed. Peter Laslett (Cambridge: Cambridge University Press, 1960).

Losano, Mario G., *Der Briefwechsel zwischen Jhering und Gerber* (Elsbach: Gremer, 1984).

Lotmar, Philip, *Der Arbeitsvertrag nach dem Privatrecht des Deutschen Reiches* (Leipzig: Duncker & Humblot, 1902).

Lotmar, Philip, *Schriften zu Arbeitsrecht, Zivilrecht und Rechtsphilosophie*, ed. Joachim Rückert (Frankfurt: Keip, 1992).

Luig, Klaus, 'Usus modernus', in *Handwörterbuch zur deutschen Rechtsgeschichte*, ed. Wolfgang Stammler, 5 vols. (Berlin: Schmidt, 1964–98), V (1998).

Lupoi, Maurizio, *Alle radici del mondo giuridico europeo: Saggio storico-comparativo* (Rome: Istituto Poligrafico e Zecca dello Stato-Archivi di Stato, 1994).

Maine, Henry Sumner, *Village-Communities in the East and West* (London: John Murray, 1871).

Maine, Henry Sumner, *Ancient Law: Its Connection with the Early History of Society and its Relation to Modern Ideas*, 3rd edn. (London: John Murray, 1866; repr. New Brunswick, NJ and London, 2002).

Mannori, Luca, and Bernardo Sordi, *Storia del diritto amministrativo* (Rome and Bari: Laterza, 2001).

Martinek, Michael, ed., *100 Jahre BGB – 100 Jahre Staudinger* (Berlin: De Gruyter, 1999).

Marx, Karl, and Friedrich Engels, *Werke*, 39 vols. (Berlin: Institut für Marxismus-Leninismus, 1964–8).

Mattei, Ugo, *Il modello di common law*, 2nd edn. (Turin: Giappichelli, 2004).

Matteucci, Nicola, *Organizzazione del potere e libertà: Storia del costituzionalismo moderno*, 3rd edn. (Turin: UTET, 2001).

Mazzacane, Aldo, ed., *Korporativismus in den Südeuropäischen Diktaturen* (Frankfurt: Klostermann, 2005).

Menger, Anton, *Das Bürgerliche Recht und die besitzlosen Volksklassen. Eine Kritik des Entwurfs eines Bürgerlichen Gesetzbuchs für das Deutsche Reich* (Tübingen: H. Laupp, 1890).

Montanari, Massimo, *La fame e l'abbondanza: Storia dell'alimentazione in Europa*, 4th edn. (Rome and Bari: Laterza, 2003).

Mortati, Constantino *La Costituzione di Weimar* (Florence: Sansoni, 1946).

Naphy, William, and Andrew Spicer, *Plague, Black Death and Pestilence in Europe* (Stroud: Tempus, 2004).

Naphy, William, and Andrew Spicer, *La peste in Europa* (Bologna: Il Mulino, 2006).

Napoli, Mario, ed., *La nascita del diritto del lavoro: Il contratto di lavoro di Lodovico Barassi cent'anni dopo; novità, influssi, distanze* (Milan: V&P università, 2003).

Necker, Jacques, *Réflexions philosophiques sur l'égalité*, in *Œuvres complètes publiées par M. le baron de Staël*, 15 vols. (Paris, 1820–1), X.

Oestmann, Peter, *Rechtsvielfalt vor Gericht Rechtsanwendung und Partikularrecht im Alten Reich* (Frankfurt: Klostermann, 2002).

Orestano, Riccardo, *Introduzione allo studio del diritto romano* (Bologna: Il Mulino, 1987).

Padoa Schioppa, Antonio, *Saggi di storia del diritto commerciale* (Milan: Edizioni universitarie di lettere, economia, diritto, 1992).

Padoa Schioppa, Antonio, *Il diritto nella storia d'Europa: Il Medioevo* (Padua: CEDAM, 1995).

Passaniti, Paolo, *Storia del diritto del lavoro* (Milan: Giuffrè, 2006).

Petit, Carlos, ed., *Del ius mercatorum al derecho mercantil: Tercer seminario de historia del derecho privado* (Madrid: Marcial Pons, 1997).

Petronio, Ugo, *La lotta per la codificazione* (Turin: Giappichelli, 2002).

Piano Mortari, Vincenzo, *Cinquecento giuridico francese: Lineamenti generali* (Naples: Liguori, 1990).

Piergiovanni, Vito, ed., *From 'Lex mercatoria' to Commercial Law* (Berlin: Duncker & Humblot, 2005).

Piovani, Pietro, *Giusnaturalismo ed etica moderna* (Bari: Laterza, 1961).

Pocock, J. G. A., *The Ancient Constitution and the Feudal Law: A Study of English Historical Thought in the Seventeenth Century*, 2nd edn. (Cambridge: Cambridge University Press, 1987).

Preuss, Hugo, *Gemeinde, Staat, Reich als Gebietskörperschaften: Versuch einer deutschen Staatskonstruktion auf Grundlager der Genossenschaftstheorie* (Aalen: Scientia, 1964).

Quaglioni, Diego, *La giustizia nel Medioevo e nella prima età moderna* (Bologna: Il Mulino, 2004).

Rabel, Ernst, 'Aufgabe und Notwendigkeit der Rechtsvergleichung', in *Gesammelte Aufsätze*, 4 vols. (Tübingen: Mohr, 1965–71), I: *Arbeiten zum Privatrecht, 1907–1930* (1965).

Rials, Stéphane, ed., *La Déclaration des droits de l'homme et du citoyen* (Paris: Hachette, 1988).

Rigaudière, Albert, Olivier Guillot and Yves Sassier, *Pouvoirs et institutions dans la France médiévale*, 2 vols. (Paris: Colin, 1994–2004), II: Albert Rigaudière, *Des temps féodaux aux temps de l'État* (2004).

Ritter, Gerhard, *Der Sozialstaat: Entstehung und Entwicklung im internationalen Vergleich* (Munich: Oldenbourg, 1989).

Ritter, Gerhard, *Storia dello stato sociale*, trans. Lorenzo Gaeta and Paola Carnevale (Rome and Bari: Laterza, 1996).

Romano, Santi, *El ordenamiento jurídico*, trans. Sebastián Martín-Retortillo (Madrid: Instituto de estudios políticos, 1963).

Romano, Santi, *Lo stato moderno e la sua crisi: Saggi di diritto costituzionale* (Milan: Giuffrè, 1969).

Romano, Santi, *L'Ordre juridique*, trans. Lucien François and Pierre Gothot (Paris: Dalloz, 1975a).

Romano, Santi, *Die Rechtsordnung*, trans. Roman Schnur (Berlin: Duncker & Humblot, 1975b).

Romano, Santi, *L'ordinamento giuridico*, 3rd edn. (Florence: Sansoni, 1977).

Rosanvallon, Pierre, *Le Peuple introuvable: Histoire de la représentation démocratique en France* (Paris: Gallimard, 1998).

Rosanvallon, Pierre, *Le Modèle politique français: La Société civile contre le jacobinisme de 1789 à nos jours* (Paris: Seuil, 2004).

Rosanvallon, Pierre, *Il popolo introvabile: Storia della rappresentanza democratica in Francia*, trans. Andrea De Ritis (Bologna: Il Mulino, 2005).

Rossi, Pellegrino, *Mélanges d'économie politique, d'histoire et de philosophie*, 2 vols. (Paris: Guillaumin, 1867).

Rousseau, Jean-Jacques, *Œuvres complètes*, ed. Michel Launay, 3 vols. (Paris: Seuil, 1971).

Royal Netherland Academy of Arts and Sciences, *The World of Hugo Grotius (1583–1645)* (Amsterdam: APA-Holland University Press, 1984).

Saleilles, Raymond, 'Étude sur la théorie générale des obligations d'après le projet de Code Civil Allemand', *Bulletin de la Société de législation comparée*, 18 (1888–9a).

Saleilles, Raymond, 'Étude sur les sources de l'obligation dans le project de Code Civil pour l'Empire d'Allemagne', *Bulletin de la Société de législation comparée*, 18 (1888–9b).

Saleilles, Raymond, 'Conception et objet de la science du droit comparé', *Bulletin de la Société de législation comparée*, 29 (1900).

Saleilles, Raymond, *De la déclaration de volonté: Contribution à l'étude de l'acte juridique dans le Code Civil allemand (art. 116–44)* (Paris: Pichon, 1901).

Saleilles, Raymond, *Introduction à l'étude du droit civil alleman* (Paris: Pichon, 1904).

Saleilles, Raymond, *De la personnalité juridique: Histoire et théories, vingt-cinq leçons d'introduction à un cours de droit civil comparé sur les personnes juridiques* (Paris: Rousseau, 1910).

Santarelli, Umberto, *Mercanti e società tra mercanti*, 3rd edn. (Turin: Giapichelli, 1998).

Sasso, Gennaro, *Antologia degli scritti politici di Immanuel Kant*, 2nd edn. (Bologna: Il Mulino, 1977).

Savigny, Friedrich Carl von, *Sistema del diritto romano attuale*, trans. Vittorio Scajola, 8 vols. (Turin: UTET, 1886–98).

Savigny, Friedrich Carl von, *System des Heutigen römischen Rechts*, 8 vols. (Aalen: Scientia, 1981).

Savigny, Friedrich Carl von, and Anton Thibaut, *La polemica sulla codificazione*, ed. G. Marini (Naples: Edizioni scientifiche italiane, 1982).

Schäffle, Albert, *Bau und Leben des sozialen Körpers* (Tübingen: Lauppschen, 1875–8).

Schäffle, Albert, *Struttura e vita del corpo sociale*, trans. Eusebio Lodovico, 2 vols. (Turin: UTET, 1881).

Schmoeckel, Mathias, *Auf der Suche nach der verlorenen Ordnung: 2000 Jahre Recht in Europa, ein Überblick* (Cologne: Böhlau, 2005).

Schneidmüller, Bernd, and Stefan Weinfurter, eds., *Ordnungskonfigurationen im hohen Mittelalter* (Ostfildern: Thorbecke, 2006).

Schnerb-Lièvre, Marion, ed., *Le Songe du Vergier* (Paris: Centre national de la recherche scientifique, 1982).

Schröder, Jan, *Recht als Wissenschaft: Geschichte der juristischen Methode vom Humanismus bis zur historischen Schule* (Munich: Beck, 2001).

Schröder, Jan, 'Deutscher usus modernus', in *Der Neue Pauly: Enzylopädie der Antike*, ed. Hubert Cancik, 16 vols. (Stuttgart: Metzler, 1996–2003), XIII (1999).

Schubert, Werner, ed., *200 Jahre Code civil: Die napoleonische Kodifikation in Deutschland und Europa* (Cologne: Böhlau, 2005).

Società italiana di storia del diritto, *Diritto generale e diritti particolari nell'esperienza storica* (Rome: Fondazione Sergio Mochi Onory per la storia del diritto italiano, 2001).

Société des Études Législatives, *Le Livre du Centenaire: Code Civil, 1804–1904* (Paris: Dalloz, 2004).

Société Jean Bodin pour l'histoire comparative des institutions, *La Coutume* (Brussels: De Boeck-Wesmael, 1990).

Stolleis, Michael, ed., *Rechtsgeschichte im Nationalsozialismus: Beiträge zur Geschichte einer Disziplin* (Tübingen: Mohr, 1989).

Stolleis, Michael, *Geschichte des öffentliche Rechts in Deutschland*, 3 vols. (Munich: Beck, 1988–99), II: *1800–1914, Staatsrechtslehre und Verwaltungswissenschaft* (1992).

Stolleis, Michael, *Geschichte des öffentliche Rechts in Deutschland*, 3 vols. (Munich: Beck, 1988–99), III: *Staats- und Verwaltungsrechtswissenschaft in Republik und Diktatur* (1999).

Stolleis, Michael, *Recht im Unrecht: Studien zur Geschichte im Nationalsozialismus*, 2nd edn. (Frankfurt am Main: Suhrkamp, 2006).

Tomas y Valiente, Francisco, *Manual de historia del derecho español* (Madrid: Tecnos, 1983).

Toniolo, Giuseppe, *Democrazia cristiana: Istituti e forme*, 2 vols. (Vatican City: Studium, 1951).

Treves, Claudio, 'Socialismo e diritto civile', *Critica sociale*, 3 (1894).

Varano, Vincenzo, and Vittoria Barsotti, *La tradizione giuridica occidentale*, 2 vols. (Turin: Giappichelli, 2002–3), I: *Testo e materiali per un confronto civil law–common law* (2002).

Verri, Pietro, *Memorie storiche sulla economia pubblica dello Stato di Milano*, Scrittori classici di economia politica, parte moderna, 50 vols. (Milan: Destefanis, 1803–16; repr. Rome: Ateneo & Bizzarri, 1966), XVII (1804).

Viano, Carlo Alberto, *Il pensiero politico di Locke* (Rome and Bari: Laterza, 1997).

Villey, Michel, *La Formation de la pensée juridique moderne*, ed. Stephanie Rials (Paris: Presses universitaires de France, 2003).

Villey, Michel, *La formazione del pensiero giuridico moderno*, trans. Rosabruna D'Ettorre and Francesco D'Agostino (Milan: Jaca Book, 2007).

Viora, Mario Enrico, *Consolidazioni e codificazioni: Contributo alla storia della codificazione*, 3rd edn. (Turin: Giappichelli, 1990).

Wieacker, Franz, *Privatrechtsgeschichte der Neuzeit unter besonderer Berücksichtigung der deutschen Entwicklung*, 2nd edn. (Göttingen: Vandenhoek & Ruprecht, 1967).

Wieacker, Franz, *Industriegesellschaft und Privatrechtsordnung* (Frankfurt: Fischer Athenäum Taschenbuch, 1974).

Wieacker, Franz, *Storia del diritto privato moderno con particolare riguardo alla Germania*, trans. Umberto Santarelli, 2 vols. (Milan: Giuffrè, 1980).

Wieacker, Franz, *Diritto privato e società industriale*, trans. Gianfranco Liberati (Naples: Edizioni scientifche italiane, 1983).

Zecchino, Orsino, ed., *Federico II: Enciclopedia Fridericiana*, 2 vols. (Rome: Istituto dell'Enciclopedia italiana, 2005).

Zorzi, Andrea, ed., *La civiltà comunale italiana nella storiografia internazionale* (Florence: Florence University Press, 2008).

Zweiger, Konrad, and Hein Kötz, *Einführung in die Rechtsvergleichung auf dem Gebite des Privatrechts*, 2 vols. (Tübingen: Mohr, 1996), I: *Grundlagen.*

Zweiger, Konrad, and Hein Kötz, *Introduction to Comparative Law*, trans. Tony Weir (Oxford: Clarendon Press, 1998a).

Zweiger, Konrad, and Hein Kötz, *Introduzione al diritto comparato*, trans. Barbara Pozzo and Estella Cigna, 2 vols. (Milan: Giuffrè, 1992–8), I: *Principi fondamentali*, 2nd edn. (1998b).

Index

Printed and bound by CPI Group (UK) Ltd, Croydon, CR0 4YY

15/09/2026

14962390-0001